AF595331

The Role of Sacrifice in the Secular Age

The Role of Sacrifice in the Secular Age

Editors

Javier Gil-Gimeno
Josetxo Beriain
Celso Sánchez Capdequí

MDPI • Basel • Beijing • Wuhan • Barcelona • Belgrade • Manchester • Tokyo • Cluj • Tianjin

Editors

Javier Gil-Gimeno
Public University of Navarra
Spain

Josetxo Beriain
Public University of Navarra
Spain

Celso Sánchez Capdequí
Public University of Navarra
Spain

Editorial Office
MDPI
St. Alban-Anlage 66
4052 Basel, Switzerland

This is a reprint of articles from the Special Issue published online in the open access journal *Religions* (ISSN 2077-1444) (available at: https://www.mdpi.com/journal/religions/special_issues/sac).

For citation purposes, cite each article independently as indicated on the article page online and as indicated below:

LastName, A.A.; LastName, B.B.; LastName, C.C. Article Title. *Journal Name* **Year**, *Volume Number*, Page Range.

ISBN 978-3-0365-2075-9 (Hbk)
ISBN 978-3-0365-2076-6 (PDF)

Contents

About the Editors

Javier Gil-Gimeno (PhD Sociology) Lecturer in the Department of Sociology and Social Work from Public University of Navarra (Spain). Member of I-Communitas. Institute for Advanced Social Research (UPNA). Their fields of interest and research are: Sociology of Religion, Sociological Theory, Cultural Sociology, Modernity and the Secular-Religious forms.

Josetxo Beriain (PhD Sociology) Professor in the Department of Sociology and Social Work from Public University of Navarra (Spain). Member of I-Communitas. Institute for Advanced Social Research (UPNA). Their fields of interest and research are: Sociology of Religion, Sociological Theory, Cultural Theory, Sociology of Time, Modernity.

Celso Sánchez Capdequí (PhD Sociology) Professor in the Department of Sociology and Social Work from Public University of Navarra (Spain). Member of I-Communitas. Institute for Advanced Social Research (UPNA). Their fields of interest and research are: Sociology of Religion, Sociological Theory, Social Imaginary and Creativity.

Preface to "The Role of Sacrifice in the Secular Age"

Dear Colleagues,

Sacrifice was one of the main features of agro-pastoral societies (Hénaff, 2010). In this type of society, sacrifice acts as a 'total social fact' (Mauss, 1979), like a primordial dimension of social life that, as such, reveals some of the main properties of this way of articulating reality and existence. In this context, the religious system was responsible for social order, and sacrifice acted as the method of communication between profane and sacred realms (Mauss, 1979). These realms, which were originally integrated into what we know as 'animistic religion' (Tylor, 1981) or 'primitive religion' (Bellah, 1969), were differentiated from each other at the same time that human beings started to develop 'second order thinking' (Elkana, 1986) or rational thinking. Here, we can clearly establish a link between the axial age (Jaspers, 1994; Eisenstadt, 1986) and 'the age of sacrifice' (Hénnaf, 2010). The role played by sacrifice has been studied by distinguished authors such as William Roberston Smith (1972), Émile Durkheim (1982), Marcel Mauss (1970; 1979), and, more recently, René Girard (1995; 2012), Marcel Hénaff (2010), and Guy Stroumsa (2012), amongst others.

We would be making a great mistake if we were to assume that sacrifice performs a similar function in current societies as in the past. As we know, societies change through time. This implies that a social fact will present several faces depending on the context in which we analyze it, and depending on the influence of the different hegemonic social forces in dispute. In its evolution, sacrifice has been necessarily affected by these dynamics of change. These have caused a transition from the imaginary focus on the religious sphere to another modern and secular one. In this transition, the role played by several "social engines" is of great importance; these include functional differentiation (Spencer, 1947; Durkheim, 1987; Luhmann, 1998, Parsons, 1977), individualization (Beck, Giddens, Lash, 1997; Bauman, 2002), secularization (Taylor, 2014; Martin, 1969; Casanova, 2012), the disenchantment of the world (Weber, 1979), acceleration (Koselleck, 2003, Rosa, 2016), and re-entchantment and re-fusion (Alexander, 2017).

We would be making another great mistake if we were to consider sacrifice as only being able to perform the role of a 'total social fact'. That is to say, either it performs this role or nothing else will do it. According to Merlin Donald (1991), social facts (evolution in his own terms) do not appear and disappear as if by magic. We witness an endless reshaping of the role that they actually represent or can represent. This paper is concerned with the social mainstream and with the values around the hegemonic institutions and social movements which are constructed in each society. In the same way, in modernity (as well as in postmodernity), it is very difficult to find 'total social facts' due to (among other things) the fragmentation of individual and collective experiences, to multiple belongings, and to functional differentiation processes.

The focus of this Special Issue is the analysis of the role played by sacrifice in complex secular and modern societies, in which, the concept of 'emotional self-restriction' (Freud, 201; Elias, 2009), as a keystone of civilization, has collapsed. Today, the old idea of sacrifice is superseded by the idea of 'useless sacrifice' (Duvignaud, 1997), not because the logic of excess carried by sacrifice is opposite to the capitalistic idea of efficacy, but mainly because the contemporary actor is far away from any ideas of containment, restraint, or control. At the base of current civilizations, 'instinctive sacrifice' is not yet the rule. We could be closer to a new version of the 'intellectual sacrifice' (Weber, 2004). The weakening of the forces of transcendence (Reckwitz, 2012) in the secular age sets up spaces of 'symbolic exchange' (Baudrillard, 1980), which play the articulator role in our hyperfragmented society. In this context, the idea of compensatory loss remains present in current wars and migratory conflicts, in the economic life of unregulated capitalism, in the new imperative of corporal beauty, in global sports competitions, and so on. All of these are contexts, current contexts, where sacrifice plays a substantive role for understanding our age.

In Merlin Donald's terms of "evolutive evolution" (1991) and with the force that drives the dynamics of change through all societies, we understand that sacrifice performs a role in current societies, but a role in which its meaning as well as its function have already changed. The aim of this Special Issue is to analyze and explain what this role is, studying some of the different social faces that it presents. Our hypothesis is radically sociological, because we understand that different dynamics of change have exerted a transformative influence over sacrifice.

In achieving this purpose, we build one structure divided into two parts: The first is named 'contextures', in which on the one hand, we analyze the role played by sacrifice in past societies or what Hénaff called 'the age of sacrifice'; and on the other hand, we attempt to urbanize the conceptual field of current sacrifice, this is, the context where sacrifice can articulate itself today. The second part is named 'textures', in which we analyze the link between sacrifice and some of the basic elements of current social life, such as economy, the nation, religion, democracy, culture, creativity, terrorism, the body, and so on.

Javier Gil-Gimeno, Josetxo Beriain, Celso Sánchez Capdequí
Editors

MDPI

Editorial

The Role of Sacrifice in the Secular Age

Josetxo Beriain , Celso Sánchez Capdequí and Javier Gil-Gimeno *

I-Communitas, Institute for Advanced Social Research, Department of Sociology and Social Work, Public University of Navarra, 31015 Pamplona, Spain; josetxo@unavarra.es (J.B.); celso.sanchez@unavarra.es (C.S.C.)
* Correspondence: fcojavier.gil@unavarra.es

Citation: Beriain, Josetxo, Celso Sánchez Capdequí, and Javier Gil-Gimeno. 2021. The Role of Sacrifice in the Secular Age. *Religions* 12: 722. https://doi.org/10.3390/rel12090722

Received: 30 August 2021
Accepted: 31 August 2021
Published: 3 September 2021

Publisher's Note: MDPI stays neutral with regard to jurisdictional claims in published maps and institutional affiliations.

Sacrifice was one of the main features of agro-pastoral societies. In this kind of society, sacrifice acts as a 'total social fact', like a primordial dimension of social life that, as such, shows up some of the main properties of this way for articulating reality and existence. In this context, the religious system was responsible for social order and sacrificial acts as the method of communication between profane and sacred realms. These realms, which originally were integrated into what we know as 'animistic religion' or 'primitive religion', were differentiated from each other around the time at the same time that human beings started to develop the 'second order thinking' or rational thinking. Here, we can clearly establish a link between what Karl Jaspers called the 'axial age' and 'the age of sacrifice', in the terms developed by Marcel Hénnaf. This kind of role played by sacrifice has been studied by distinguished authors such as William Roberston Smith, Émile Durkheim, Marcel Mauss, or, more recently, René Girard, Marcel Hénaff and Guy Stroumsa, among others.

It would be a great mistake to assume that sacrifice performs the same function in current societies that it fulfilled in the past; as we know, societies change with time. This implies that a social fact will present several faces depending on the context that we analyze and depending on the influence of the different and hegemonic social forces in dispute. Through their evolution, sacrifice has necessarily been affected by these dynamics of change. These have caused the transition from one imaginary focus on the religious sphere to another, which is modern and secular. In this transition, it is really important to consider the role played by several "social engines", such as functional differentiation, individualization, secularization, the disenchantment of the world, acceleration, re-entchantment and re-fusion.

It would be another great mistake to assume that sacrifice can only perform the role of a 'total social fact'. That is to say, either it performs this role, or no one else will do it. According to Merlin Donald, social facts (evolution in his own terms) do not appear and disappear as if by magic. We witness an endless reshaping of the role that they actually represent or can represent, and this paper is very much connected with the social mainstreams, as well as with the values around hegemonic institutions and social movements which are constructed in each society. In the same way, in modernity (and in post-modernity too), it is very difficult to find 'total social facts', due to (among other things) the fragmentation of individual and collective experiences, to the multiple belongings, to functional differentiation processes.

In Merlin Donald's terms of "evolutive evolution" (1991) and with the strength that exerts the dynamics of change around the whole society, we understand that sacrifice performs a role in current societies, but a role in which its meaning as well as its function have already changed. We wanted to explain what this role is and to study some of the different social faces that it presents.

In achieving this purpose, this special issue includes seven papers that we are briefly going to introduce the following:

In their work, entitled "The Endless Metamorphoses of Sacrifice and its Clashing Narratives", Josetxo Beriain (I-Communitas, Institute for Advanced Social Research, Public

University of Navarra) develops a fourfold task, which: 1. Provides an affirmative genealogy that shed light on the different forms taken by sacrifice, the origins of its various conceptual layers and the various social practices from which they come: 2. Analyzes the initial conceptual layer proposed by Marcel Mauss and Henri Hubert and followed by Marcel Hénaff based on farming societies; 3. Analyzes the rise of the anti-sacrificial narrative and its main landmarks, the problems of victims and the responses given by René Girard and Talcott Parsons; and 4. Analyzes the dynamic tension between the tragic-apocalyptic narrative and the defensive-progressive narrative in modern times, and the main landmarks of each one.

In "A Maximal Understanding of Sacrifice: Bataille, Richard Wagner, Pilgrimage and the Bayreuth Festival", Philip Smith (Yale University) and Florian Stoll (Leipzig University; Bayreuth University) call for a broad conception of sacrifice to be developed as a resource for cultural sociology. They argue the term was framed too narrowly in the classical work of Hubert and Mauss. The later approach of Bataille permits a maximal understanding of sacrifice as non-utilitarian expenditures of money, energy, passion and effort directed towards the experience of transcendence. From this perspective, pilgrimage can be understood as a specific modality of sacrificial activity. This paper applies this understanding to the annual Bayreuth "Wagner" Festival in Germany, while the article traces sacrificial expenditures at the level of individual festival attendees. These include financial costs, arduous travel, dedicated research of the artworks, and disciplines of the body. Some are lucky enough to experience transcendence in the form of deep emotional experience, and a sense of contact with sacred spaces and forces.

In "Metamorphosis of the Sacrificial Victimization Imaginary Profile within the Framework of Late Modern Societies", Ángel Enrique Carretero Pasín (University of Santiago de Compostela) aims to analyze the imaginary profile of the emerging sacrificial victim in late modern societies. For doing this, Carretero develops a work based on three steps: 1. He analyzes the nature and the functionality of an anthropological structure linked to a rituality of sacrificial victimization surviving in the historical course of western societies; 2. He studies the characterization of the imaginary paradigm of sacrificial victimization crystallized in modernity in contrast to the dominant one in the *Old Regime*. 3. The generalization of a climate of violence that transforms any individual into a potential victim of sacrifice is analyzed as the unique morphology of the imaginary of sacrificial victimization that emerged in late modern societies.

In "The Dark God: The Sacrifice of Sacrifice", Joseba Zulaika (Center for Basque Studies, Reno University) draws from the Frazerian question of murder turned into ritual sacrifice for implementing it to the basque case. The work addresses such a "sacrificial crisis" in the experience of their own Basque generation. He argues that the crisis regarding sacrifice is pivotal for understanding it. In achieving this, Zulaika expands the notion of "sacrifice" from my initial approach of ethnographic parallels towards a more subjective and psychoanalytical perspective. For him, the motivation behind the basque violence (focused in ETA terrorist group) was originally and fundamentally *sacrificial*; when it finally stopped in 2011, many of those invested in the violence, actors as well as supporters, felt destitute and had to remodel their political identity. The argument of the paper is that the dismantling of sacrifice as its nuclear premise—the sacrifice of sacrifice—was a major obstacle, stopping the violence from coming to an end.

In, "Trauma and Sacrifice in Divided Communities: The Sacralisation of the Victims of Terrorism in Spain", Eliana Alemán (Public University of Navarra) and José M. Pérez-Agote (I-Communitas, Institute for Advanced Social Research, Public University of Navarra) aim to show that the sacrificial status of the victims of acts of terrorism, such as the 2004 Madrid train bombings ("11-M") and ETA (Basque Homeland and Liberty) attacks in Spain, is determined by how it is interpreted by the communities affected and the manner in which it is ritually elaborated a posteriori by society and institutionalised by the state. The paper also explores the way in which the sacralisation of the victim is used in socially and politically divided societies to establish the limits of the pure and the impure in defining the

"Us", which is a subject of dispute. To demonstrate this, they first describe two traumatic events of particular social and political significance (the case of Miguel Ángel Blanco and the 2004 Madrid train bombings). Secondly, they analyse different manifestations of the institutional discourse regarding victims in Spain, examining their representation in legislation, in public demonstrations by associations of victims of terrorism and in commemorative "performances" staged in Spain. The conclusion is that in societies such as Spain's, where there exists a polarisation of the definition of the "Us", the success of cultural and institutional performances oriented towards reparation of the terrorist trauma is precarious. Consequently, the validity of the post-sacrificial narrative centring on the sacred value of human life is ephemeral and thus fails to displace sacrificial narratives in which particularist definitions of the sacred "Us" predominate.

In "Debt and Sacrifice: The Role of Scapegoats in the Economic Crises", Luis Enrique Alonso (Autonomous University of Madrid) and Carlos J. Fernández Rodríguez (Autonomous University of Madrid) assert that one of the spaces where sacrifice actually performs a critical role is the realm of modern economy, particularly in the event of a financial crisis. They analyze how the hegemonic narrative has clear sacrificial aspects. Such crises represent situations defined by an outrageous symbolic violence in which social and economic relations experience drastic transformations, and their victims end up suffering personal bankruptcy, indebtedness, lower standards of living or poverty. Crises show the flagrant domination present in social relationships: this is proven in the way crises evolve, when more and more social groups marred by a growing vulnerability are sacrificed to appease financial markets.

In "The Persistence of Sacrifice as Self-Sacrifice and Its Contemporary Embodiment in the 9/11 Rescuers and COVID-19 Healthcare Professionals", Javier Gil-Gimeno (I-Communitas. Institute for Advanced Social Research, Public University of Navarra) and Celso Sánchez Capdequí (I-Communitas, Institute for Advanced Social Research, Public University of Navarra) analyze the persistence of sacrifice as self-sacrifice in contemporary societies. In order to reach this goal they develop a work in four steps: 1. They discuss how in the Axial Age (800–200 B.C.E.) an understanding of sacrifice as ritual worship or a ritual practice that involves the immolation of a victim became less prevalent and a new understanding of sacrifice emerges. This new notion of sacrifice focuses on individual relinquishment and gift exchange, that is, on a person relinquishing him/herself as a gift that is given in an exchange relationship for protecting a greater good. 2. They analyze how this new sacrifice formula led people to conceptualize sacrifice as a project or as something that persons could intentionally embrace. 3. They attend to the secularization of sacrifice, not in the sense of a de-sacralization of this phenomenon but in the way of sacralization of the mundane realm and mundane things, such as intentional self-sacrificial acts, in social contexts where there is religious pluralism. 4. They study the sacredness of the person as a clear type of secular religiosity that develops self-sacrificial forms. Two of these self-sacrificial forms are the actions of 9/11 rescuers and COVID-19 healthcare professionals. A short analysis of both serves to illustrate how self-sacrifice is embodied in contemporary societies.

In essence, sacrifice persists in modern and secular societies in an 'evolutive evolutionary' way. As point by authors like Merlin Donald or Robert N. Bellah, nothing is lost in social evolution. Previous sacrificial forms remains and live together with the new form that this phenomena acquires. This scenario provokes dynamic tensions but also a great pluralism or diversity of forms that sacrifice can develop: sacrificial and antisacrificial narratives, individual or comunal, religious or secular, and so on. The papers conform this Special Issue are a clear example of this sacrificial pluralism.

We do not finish this editorial without express our gratitude to all the persons and institutions that have made posible this work, particularly the authors that have took part in this work, and the institutions they represent; to *Religions* journal, above all Macy Zong; Both the Research Institute I-Communitas, Institute for Advanced Social Research and the Research Group 'Cambios Sociales' from Public University of Navarra, for helping us to fund an important part of the translations; to the Research Project: CSO2017-85052-R "Las variedades de la experiencia creativa y modelos de sociedad" funding by Ministerio de Economía, Industria y Competitividad from Spain, for supporting the cost of this special issue as a key outcome of the research we have developed during the years 2018–2020.

Finally, the authors of this special issue entitled: "The Role of Sacrifice in the Secular Age" certify that they have no conflict of interest in the subject matter or materials discussed in this manuscript.

Funding: The APC was funded by Research Project: CSO2017-85052-R "Las variedades de la experiencia creativa y modelos de sociedad" funding by Ministerio de Economía, Industria y Competitividad from Spain. Translations were partly funded by Research Institute I-Communitas, Institute for Advanced Social Research and the Research Group 'Cambios Sociales' from Public University of Navarra.

Conflicts of Interest: The authors declare no conflict of interest.

Article

The Endless Metamorphoses of Sacrifice and Its Clashing Narratives

Josetxo Beriain

I-Communitas, Institute for Advanced Social Research, Public University of Navarre, 31006 Pamplona, Spain; josetxo@unavarra.es

Received: 19 November 2020; Accepted: 16 December 2020; Published: 19 December 2020

Abstract: This paper sets out (1) to provide an affirmative genealogy that shed light on the different forms taken by sacrifice, the origins of its various conceptual layers and the various social practices from which they come; (2) to analyze the initial conceptual layer proposed by Marcel Mauss and Henri Hubert and followed by Marcel Hénaff based on farming societies; (3) to analyze the rise of the anti-sacrificial narrative and its main landmarks, the problems of victims and the responses given by René Girard and Talcott Parsons; and finally (4) to analyze the dynamic tension between the tragic-apocalyptic narrative and the defensive-progressive narrative in modern times, and the main landmarks of each one.

Keywords: sacrifice; gift; victim; post-heroic; sacralization of the person

---

1. Introduction: Distinction between Offerings and Sacrifice

There are several erroneous prior assumptions that must be avoided when addressing the meaning and different expressions of sacrifice. The first is the idea that sacrifice fits into a purely taxonomic and classificatory approach obsessed with classifying it into different types without first defining it. Certain late 19th century British anthropologists (Tyler, Frazer) strove to do this. The second idea is the reducing of sacrifice to a mere expression of primitive barbarism as opposed to civilization. The third is that this sacrificial expression disappears within the format of a markedly teleological socio-anthropological conjecture, i.e., that the more modern and civilized a society is, the less the culture of sacrifice is present in it. Here I demonstrate that these assumptions are empirically false.

There is a large body of documentation on the history of religions, anthropology and sociology that includes tales of offerings and sacrifices made all over the world by groups of hunter-gatherers and, especially, in pastoral societies, i.e., among farmers. The first clue to profiling the concept of sacrifice can be found in certain Indo-European languages, as indicated by Emile Benveniste (1969, pp. 223–26, 187–88) and Joseph Henninger (1995, p. 544 et seq.). In Vedic Sanskrit there are other roots that also refer to the idea of sacrifice: *Hav- juhoti* "to sacrifice", *hotar* "sacrificial priest", *hotra* "sacrifice", *Agni-hotra* "sacrifice to the god Agni". Our term "sacrifice" comes from the Latin *sacrificium* (*sacer*, "holy"; *facere*, "to do"); it can also be understood as the act of sanctifying or consecrating an object. "*Offering*" is used as a synonym (or as a more inclusive category of which sacrifice is a subdivision) and means the presentation of a gift. "Offering" has its roots in the verb *offerre*, "to offer, present"; the verb yields the noun *oblatio*. The German word *Opfer* ("victim") also comes from *offerre*, but some experts also point to an etymological relationship with *operari* ("to act" or "to carry out"), thus evoking the idea of sacred action.

But although "offering" and "sacrifice" are related etymologically, in the ambit of ritual actions they are not the same thing. Jan van Baal introduces a major guiding distinction under which "I call an offering every act of presenting something to a supernatural being, a sacrifice an offering accompanied by the ritual killing of the object of the offering" (van Baal 1976, p. 161). A precedent for this can be found in a definition in William Robertson-Smith's contribution to the 1886 edition of the *Encyclopaedia Britannica*, where he distinguishes clearly between two types of sacrificial offering: "*Hostia honoraria* refers to the case in which the deity accepts a gift, while *hostia piacularis* refers to that in which the deity demands a life (the death of a living being, animal or human)" (Robertson-Smith 1886, p. 132).

2. Social Origins of Sacrifice

This distinction is crucial not just in determining the social origins of sacrifice but also in delimiting the concept itself (Beriain 2017). It was Marcel Mauss who first observed, back in 1924, that over 10,000 years ago, before the rise of agriculture and livestock farming in the Neolithic, there was already a kind of ceremonial exchange of gifts (offerings-presents-gifts) (Mauss [1924] 1971, pp. 155–269) between a sacralized nature and human beings (groups of hunter-gatherers), whose way of life had existed for two and a half million years. The invention of agriculture (Leakey 1982, 1994) gave rise to a new type of exchange between gods and human beings (represented by groups of agricultural and livestock farmers): sacrifice. More recently, Marcel Hénaff ([2002] 2010, p. 164 et seq.), basing his arguments on, among other things, field work by Roberte Hamayon (1990, p. 375) involving Siberian hunting societies, states that between humans and animals there was an egalitarian relationship based on an alliance and the exchange of gifts. Kinship structures, as the core principle of social organization in any segmented society, extend to all living beings and creatures, rivers and mountains. All the ambits that make up reality are related. There is no ontological superiority through which any one of these elements stands out from the rest. An animal is a part of the nature that we take-hunt-receive, but for that we must offer-give back something in return. Life is a gift that we receive and do not produce ourselves, but for which we must offer-give back something somehow. Life itself is a gift. The sacredness of life manifests itself in its circulation through all areas of the real. Among hunter-gatherers, nature was not seen as something hierarchically lower than the divine, and nor was the divine considered as something separate above it. On the contrary: nature was sacralized and the natural world was supernatural. Spirits were not the abstract, deified figures that they were to become in later cultures but rather served as "magical potentialities" (Mauss 1971, pp. 122–33)—*mana, wakan, orenda, manitu, daimon*—that moved from one plane of reality to another. No-one was the sole owner of that "magical potentiality": it was rather something that circulated, that must circulate, between nature and society: it was life as a gift, as a "total social fact" (Mauss 1971, pp. 157, 203) that jointly implied all instances of what was real. Gift-giving was a device that linked all members of a group (men and women, ordinary members and chiefs), that linked humans with objects, the living with the dead, humans with their *daimons*. This makes up a symbolic whole that links together the various parts of what is real (Mauss [1924] 1971, p. 195). This is a magical world accounted for in mythologies where reality is expressed symbolically[1].

Roberte Hamayon observes that in the world of farmers and shepherds, unlike that of hunter-gatherers, "the supernatural (*la surnature*) becomes vertical, and with that relationships become hierarchical, humans no longer treat it as an equal. They feel that their commitment to the supernatural is no longer based on a position of equality but on a dependence relationship. Human beings venerate and implore their ancestors, who are located above them in both time and space, because they believe that they reside in the mountains overlooking their pastures, and they hope for their protection" (Hamayon 1990, p. 737). Marcel Hénaff ([2002] 2010, p. 171) argues that what has happened is that the alliance-type relationships characteristic of hunter-gatherers were replaced by relationships based on kinship and lineage, so gifts were no longer bestowed by nature but by ancestors and heirs. Animals and plants were no longer free beings that lived in nature but were *produced* (as livestock raised, fenced in and branded, owned by someone, and as crops and plants grown by peasants) by human hand.

---

1 Merlin Donald (1990) established the main outlines of an evolutionary process that began with *Homo Erectus* and the *mimetic culture* two million years ago, continued with the *mimetic/symbiotic culture* described here as from around 250,000 years ago and led to the *theoretical culture* that peaked in the Axial Age.

When wild animals belonged to no-one but nature they could not be sacrificed, because they were not property. However, once they were *appropriated* by human beings they could be sacrificed to give back to the gods (no longer to *daimon* spirits) what had been received from them. This is where *offering* was transformed into *sacrifice*. Marcel Hénnaf holds that in hunter-gatherer societies animals were the incarnation of a *daimon*-spirit ally, but in farming societies they were split into two: domesticated animals (which were friendly and sacrificeable) and wild animals (which no longer represented deities, were not allies of human beings and were not sacrificeable) (Hénaff [2002] 2010, p. 173).

What reasons can be called on to interpret this transition from an alliance-exchange-giving between living beings through the *daimon-mana* to sacrifice as mediation between humans and gods? I believe that there are at least two fundamental reasons: the first is the rise of the altar as a place of sacrifice (Harrison 1912, p. 147), which displaces eating together based on the idea of everyone being equal. In sacrifices on an altar three distinct figures emerge: the offeror, the sacrificial victim which is offered up and the god, separate from mortals, to whom the offering is made. This idea is analyzed in greater detail in the next section. The second reason is that religious value ceases to reside in the alliance-exchange-giving and its symbolic expression in the magical potentiality of the *Daimon-mana*, and instead begins to take the form of a distinction between the transcendent world and the immanent world (Weber 1978, p. 412; Eisenstadt 1986, pp. 1–29; Schwartz 1975, vol. 2, pp. 3–4), as a particular and rather historical variant of the set of systems for the universal, dualistic classification of social reality into the sacred and the profane as postulated by Èmile Durkheim, and above all of the supramundane and inframundane areas that emerged in the Axial Age, as described by Max Weber and S. N. Eisenstadt.

3. Conceptual Origins of Sacrifice

Once the social origins of sacrifice have been established, it becomes possible to pin down its conceptual structure. William Robertson-Smith asserts (and this is one of the crucial aspects of sacrifice as a constitutive ritual) that "the victim was naturally holy, not in virtue of its sacrificial destination but because it was an animal of holy kind" (Robertson-Smith [1889] 1972, p. 390). However, there is a methodological error here: how can the victim be considered as holy *per se*, before the sacrifice, if, as stated above, in farming societies only the realm of the divine can be holy? It must be recalled that in hunter-gatherer societies the *daimon-mana-giving* circulated from one milieu to another and was not owned by any single one of them. It is in farming societies, where the transcendent world (bringer of the sacred) has been separated from the immanent/profane world, where the "*procedure consists in establishing a means of communication between the sacred and the profane worlds through the medium of a victim, that is, of a thing that in the course of the ceremony is destroyed*" (Hubert and Mauss [1898] 1981, p. 97, their italics)[2]. Contrary to what Robertson-Smith believed, Hubert and Mauss held that the victim did not necessarily come with a religious nature already perfected and a clearly defined religious nature: it is the sacrifice itself that confers this upon it (Hubert and Mauss [1898] 1981, p. 97). It is the sacrificial ritual that imparts sacredness to the victim. It is the ritual that creates sacredness: it does not exist *per se*. The sacredness of the victim does not exist as a prior idea or belief but comes about only through the ritual. The existence of this sacredness precedes the essence of/belief in it. What makes something sacred is the collective feeling that accompanies it. Only through the appropriating event of the ritual does sacredness emerge as something distinct from the profane. Through intensified action and emotional energy, the ritual not only brings about a transcendence of the world taken for granted, the everyday, profane world, to create the "*sphere of the sacred*" (Durkheim [1912] 1982,

[2] Claude Levi-Strauss defines sacrifice in similar terms as "an irreversible operation (the destruction of the victim) in order to release, on another level, an equally irreversible operation (the granting of divine grace), which is required by the fact that two "recipients" situated at different levels, have previously been brought into communication" (Levi-Straus 1964, p. 327). For more about the social origins of the guiding distinction between the sacred and the profane and its historical metamorphoses, see the paper by Beriain (2015, vol. 151, pp. 3–22).

p. 205) and, by extension, sacralizing the victim, but also unifies individuals and thus gives rise to a single collective whole. Sacrifice is the bridge that links the two worlds and at the same time the door that separates them: a door that can only be opened by ritual, through the practice of sacrifice. It is an act that requires giving and receiving, but with the intermediation of a subject that is immolated. This is an innovation in comparison with mere "offering". Humanity must give back, even if only in small measure, what it has received as a gift from the gods (no longer from nature): the victim is the device for mediation with the world of the sacred (the gods) and also the counter-gift offered up by the human world in payment of the debt of humanity to god[3]. Ultimately, sacrifice leads to an exchange of gifts and counter-gifts similar to that described by Marcel Mauss in his *Essay on the Gift* in 1924, by E. E. Evans-Pritchard ([1956] 1981, p. 326) and by Marcel Hénaff ([2002] 2010, p. 200). It is a "means of symbolic communication that jointly involves the sacred and profane worlds that make up the real" and thus form a "total social fact" (Mauss [1924] 1971, pp. 157, 203). Not only that, but "*[s]acrifice is a religious act which, through the consecration of a victim, modifies the condition of the moral person who accomplishes it*" (Hubert and Mauss [1898] 1981, p. 13, their italics).

But what elements are involved in sacrifice? The first point to take into account is that sacrifice is a rite of passage (Arnold van Gennep [1909] 1986, p. 103 et seq.), i.e., it requires that a number of rules for purification be followed, without which it is not possible to pass from the profane world to the sacred world (sacralization) and vice versa (desacralization). According to Hubert and Mauss the first step in the ritual is the "entry into the sacrifice", in which the key roles are those of the "*sacrifier* [...], *the subject to whom the benefits of sacrifice thus accrue or who undergoes its effects* [...]. This subject is sometimes an individual, sometimes a collectivity" (Hubert and Mauss [1898] 1981, pp. 10, 20 et seq.) and the *sacrificer*, the intermediary, the priest, the visible agent of consecration, who stands on the threshold of the sacred and profane worlds and represents them both at one and the same time. They are linked in him (Hubert and Mauss [1898] 1981, pp. 22–25). The next major element in the sacrificial process, according to Hubert and Mauss, is the victim. The priest, the altar and fire are also essential elements. The sacrifier comes into contact with the victim only through the priest/officiator/sacrifice, and the latter does so not directly but through the instruments provided for that purpose. The sacred and the profane may not "touch each other" directly, but must do so through the mediation of the sacrificial victim. The culmination of the ceremony comes with the death of the victim, once the spirit that inhabited the profane body of the victim enters the sacred world of divinity. The priest/sacrificer charged with taking the action has to undergo an act of purification on exit, like the "expiation of a criminal" who has killed someone. This act of destruction represents the essence of sacrifice, when the victim is separated once and for all from the profane world, sacrificed and therefore "consecrated" (Hubert and Mauss [1898] 1981, p. 35). Hence the etymology mentioned above of *sacrificium* (*sacer*, "holy"; *facere*, "to do"). The victim now becomes a creature reborn, but in the sacred world. The final element in the ritual process is the instance to which sacrifice is addressed, i.e., the divinity, which holds ontological pre-eminence in the system of classification of farming societies in which sacrifice is practised. However, just as the ritual process began with a rite of entry, it ends with rites of exit. The closest, clearest example can be found in the Roman Catholic mass, when after communion the priest cleans the chalice and wipes his hands. Once this has been done the mass is ended, the ceremonial cycle is completed and the priest utters the words of dismissal: *Ite, missa est*. Similar exit rites can be found in the *Brahmanas*.

3 However, in the Exchange of gifts studied by Mauss ([1924] 1971, pp. 155–267, 213–15), under the heading of receiving-giving-offering back, individuals must always offer more than they receive (in fact they must offer *themselves* in the exchange) in the sacrifice in which there is an exchange between humans and gods. Durkheim asserts that in reality a person "gives to sacred beings a little of what he receives from them and he receives from them all that he gives them" (Durkheim [1912] 1982, p. 317). In other words, humans always receive more than they give to the gods.

4. Archetypical Anti-Sacrificial Landmarks and the New Anti-Immolation Narrative

The end of the fascination with immolation, sacrificial ritual and victims led to a major metamorphosis in sacrifice. It lost some of its characteristic traits—those related to the act of immolation—but that does not mean that it died out, since it still maintained its core significance as a relational nexus between the sacred and the profane. Within the profane, secular world sacrifice continued to exist, as shown below. It was the narrative and the heterodox construction of symbols that arose subsequently in Hebrew monotheism that were to bring the question of ethics into the heart of sacrificial practice. There was a shift in the axis of the relevance of the divine to the human world, and especially in the idea of the sacrifice of human beings, through the mediation of virtuous men of religion (prophets) who came out of the ethical prophecies of the great Semitic tradition. The main point of this major shift in emphasis lies not so much in the point of origin of a myth in the past as in the persistence of the subsequent narrative in communities of interpretation and action. Without the unique promises of the great unknown writer of the time of exile who drew up the prophetic theodicy of suffering, misfortune, poverty, humiliation and ugliness, especially the doctrine of the Servant of Yahweh which teaches that although blameless he suffered and died voluntarily as an expiatory victim, the subsequent development of the Christian doctrine of the martyrdom of the divine Savior would not have been possible (Weber 1987, 21 and ss).

Without doubt, one of the landmarks of this narrative is the story of Abraham and the non-consummated sacrifice of his son Isaac. God tempted Abraham, saying to him "Take now thy son, thine only son Isaac, whom thou lovest, and get thee into the land of Moriah; and offer him there for a burnt offering upon one of the mountains which I will tell thee of" and just when Abraham was about to sacrifice his beloved son, God spoke to him again: "Lay not thine hand upon the lad, neither do thou any thing unto him: for now I know that thou fearest God, seeing thou hast not withheld thy son, thine only son from me". Sören Kierkegaard wrote in *Fear and Trembling* that to be absolutely responsible to God, Abraham must sacrifice his ethical duty, precisely because he loves his son unconditionally, on the altar of his unbreakable faith in God, which makes him religiously a devout believer but ethically a murderer (Kierkegaard 1976, pp. 105–6; Derrida 1995, p. 65; Gordon 1995, p. 60; Zulaika and Douglass 1996, p. 123 et seq.; Zulaika 2020). God himself, in an anti-sacrificial attitude, stays Abraham's hand from killing. The latter's exercise of faith, on the same level as his love for his son, sublimates the action of sacrifice or at least leaves it in suspension. In principle faith, trust in God, acts as a functional equivalent of sacrifice.

The Abrahamic and Kierkegaardian ambivalence of being a believer or a killer is diffused in Jesus Christ, perhaps the strongest link in the anti-immolation narrative. René Girard ([1978] 1982, p. 214) states that the heart of this shift away from sacrifice lies in the Gospel of St Matthew. In his incarnation as a human being Christ, the bearer of the divine, takes on the suffering, the pain of being cast into the world (recall the words of anguished impotence and final abandonment of Jesus on the cross, when he cried "My God, my God, why hast Thou forsaken me? (*Eli, Eli, lama Sabachtani*)". Girard proposes a re-mythologization, a shift in the narrative or sacrificial ritual, presented as a symbolic realization in which the myth of bread and wine (*metaphor)* symbolizes a new gift-giving which is materialized with the presentialization of the sacred (*sacrament*) through a humanized God (both victim and God). This sacrificing *of* God heralds the end of sacrificing *to* God. This is where the sacrifice of sacrifice itself can be said to begin (Keenan 2005, pp. 124–25). The truth of the sacrifice that is revealed in the crucifixtion, in the *kenosis* of Christ, destroys once and for all the reason for all sacrifices. This mystery is the self-humiliation, *kenosis*, of God, who descends from the infinite majesty of divinity to take the form not just of a human being as such but of a human being who is rejected, mocked and ultimately killed in the most degrading circumstances. By taking on the evils of the world, Christ announces a reconciliation with no ulterior motives and no sacrificial intermediaries. The ethics of Jesus' Sermon on the Mount are linked to an *alliance* (the third, the first being with Abraham and the second with Moses) which will excise violence from the community. In Christianity the new prevailing tone is that of love: "[f]or God so loved the world that he gave his only begotten Son, that whosoever believeth in Him should not perish but have everlasting life". However, this God loves not only the chosen people but the whole world. Salvation depends on faith and is conceived as a gift from God, while in the old testament scriptures the giving went in the opposite direction, i.e., sacrifices *to* God and obedience of his commandments (Parsons 1978, p. 271); the only condition is faith, trust in God. Not even works are ultimately a determinant. Taking Marcel Mauss's 1924 *Essay on the Gift* as a reference point for his interpretation, Talcott Parsons (1978, pp. 264–99) analyses death as a major contributor to the evolutionary enhancement of life, and thereby it becomes a significant part of the aggregate "gift of life" that all particular lives should end in death. He makes no reference to sacrificial elements, but asserts that "Death acquires a transbiological meaning because the paramount component of its meaning is the giving of life, at the end of a particular life, to God (or to man) as an expression of love for God (or man). This seems to symbolize the conception of a perpetual solidarity between the bio-human level, symbolized by the blood of Mary, and the divine level, symbolized by the blood of Christ. In the ideal Christian death, one came to participate in the blood of Christ at a new level. This is the reciprocation of God's gift to mankind through Mary" (Parsons 1978, p. 275).

In the plot of Sophocles' Greek tragedy *Antigone*, King Creon refuses to allow the body of Polynices to be buried following the rites established for heroes who have fallen in battle. His sister Antigone attempts to have him buried with honours and Creon sentences her to be buried alive. She then decides to hang herself. This tragedy has been interpreted in many different ways (Lacan 1992; Butler 2002; Irigaray 1992; Heidegger 1984, pp. 145–46), but a common thread running through those interpretations is the fact that the institutive power of sacrifice represented by Antigone competes with and questions the instituted power of sacrifice which Creon seeks symbolically to monopolize. This act of ethical appropriation on the part of Antigone entails to some degree an act of political expropriation of Creon, of his symbolic male dominant order and of the gods that he represents. Antigone appears as a woman who acts autonomously, as a pure and simple link between the human being and that of which he miraculously happens to be the bearer, the signifying cut that confers him the indomitable power of being what he is in the face of everything that may oppose him, both the gods above and those below, including death, as the possibility of absolute impossibility (Lacan 1992, p. 282).

Christianity rejects the idea of violent (ritual) sacrificial death and revolts against it, even when it wears the trappings of martyrdom. After St Paul, the idea of the sacrifice of the Son of God became intolerable unless it could be understood as a mechanism for assuring his resurrection and return to life as a spirit/*daimon* that lives among us. Sacrifices, especially blood sacrifices, were at the heart of religious activities in the farming societies of the ancient world, especially official, public religious activities, but with the law introduced by Constantinius II in the 4th century things shifted to "*sacrificiorum aboleatur insania*" (Stroumsa 2009, pp. 57–58). Guy Stroumsa states that among the Jews and other communities sacrifice was replaced by prayer (Stroumsa 2009, p. 63). The destruction of the temple at Jerusalem resulted in a major shift towards rituals without priests or blood sacrifices, where rabbis had no liturgical role. It was no longer the smell of smoke or of roasted meat typical of sacrifice that was pleasing to God. The water of ablutions and baptism has replaced the fire of sacrifice for Christians and Jews, and the soul as an interior temple has transformed rituals of purification into rituals which are ascetic rather than expiatory.

Marcel Hénaff has explored a great many post-immolation realities, focusing on the phenomenon of the "grace of God" (Hénaff [2002] 2010, chp. 7): *Kharis* in ancient Greek, *kharis* also in the Christianity of St Paul, *gratia* in Seneca and St. Augustine and *hén* in the biblical scriptures. The narrative that focuses on the ambit of grace is a new feature in regard to the ceremonial exchange of gifts and sacrificial rituals per se, given that it entails a rethinking of the social bond, with the emergence of a divine instance that unconditionally grants its grace (divine favor or friendly action) to a whole community of believers. It consists of providing a service for nothing in return. This free service, however, gives rise to acknowledgement. Generosity and acknowledgement (Benveniste 1969, vol. 1, pp. 199, 201) therefore appear as jointly involved. Grace is always something extra, over and above "what counts", what is oligatory or predictable; "it belongs to the register of the extraordinary (hence its association with the sacred)" (Pitt-Rivers 1993, p. 284). *Kharis* can be seen as designating the state of a subject (such as joy or pleasure), an attribute of an object (such as charm or beauty, hence the *Three Graces* by Raphael and Rubens) or the resulting attitude (gratitude) (Hénaff [2002] 2010, p. 245). There are several verb phrases that convey the diverse meanings of grace: *gratias agere* (to give thanks), *gratiam debere* (the duty of acknowledgement), *gratiam referre* (to do a favour) and *gratiam inire* (to earn someone's good graces).

The Hebrew equivalent of the Greek *kharis* is *hen*, which refers to the gift—the gesture of generosity–of the one God. It means a benevolent act by a person of high rank towards one of lower rank. The initial gift of Yahweh is precisely the choosing—*bahhar*—of his people from among many other peoples, as an opening, sovereign gesture confirmed through numerous divine initiatives such as the calling of Abraham, the gift of land, the exodus and the first kingdom. That gift received as a people can never be reciprocated or equaled, and it is important to recognize that impossibility, based on the strictly unconditional, transcendent, inexplicable favour of Yahweh. The chosen are linked via the bond—*berith*—that refers them to a higher, giving instance in the form of Yahweh (Hénaff [2002] 2010, pp. 253–54), so the bond is one of vassalage, protection and dependence/ subordination, in contrast with the bond of interdependence and alliance between hunter-gatherers and their *daimons*.

5. The Clash of the Sacrificial and Anti-Sacrificial Narratives in Modern Times

Societies do not always sacralize the same realities. In the first place, as seen above, societies made up of groups of hunter-gatherers sacralize nature (*la surnature*). Secondly, gods in the West/impersonal powers in the East were sacralized in the Axial Age, and their bearers were groups of farmers. A third landmark is the sacralization of a king or governor as "the highest of men and the lowest of gods, the link between the dead, the living and the immortal" (Joas 2017, p. 463). In ancient societies the fusion of sacrality and highly concentrated power gave rise to new social constellations (Erkens 2013, pp. 15–32; Joas 2017, p. 465). Royalty stems from charismatic heroism (Weber 1978, p. 875). Fourthly, in the 18th century there was a collective self-sacralization that entailed changes in the idea of the sacred governor in the Age of Absolutism ("he who sits on the divine throne and is appointed by Him". Louis XIV: *Le Roi-Soleil*). This in turn gave rise to a potential opponent of the king represented by another subject of sacralization: the people of the nation. With the "people of the nation" an "in group/outgroup" distinction emerged in a narrative created mainly by the different varieties of nationalism, which led to a form of sacralization of the "people of the nation" or "the nation" resting on "true, sacred customs" which have somehow been profaned (Akenson 1992).

5.1. The Narrative of the Sacralization of the Nation and the National Hero

The clash between the sacrificial/apocalyptic (Smith 2005) and anti-sacrificial narratives continued into the Age of Modernity. The nation, as the new "god-totem", continued the narrative of sacrifice and took over the place once held by classical divinity, creating its own altars, monuments and sacrificial, ritual commemorations.

But in the centuries that elapsed from early Christianity to the 18th century major semantic transformations took place in the narrative of sacrifice. It would not be possible to understand the culture of sacrifice in the early part of the Age of Modernity without considering those transformations.

5.1.1. The Transition to Modernity: "Pro Patria Mori"

In the blend of conceptual horizons that came with Aristotle, Christian patristics and Renaissance and Baroque humanism, there was a major secularization of categories, since the old metaphor of the marriage between bishops and their seats served to interpret relations between princes and the state: "And just as men are joined together spiritually in the spiritual body, the head of which is Christ . . . , so are men joined together morally and politically in the *res publica*, which is a body the head of which is the Prince" (Kantorowicz 2012, p. 229). In the Middle Ages, "the duty to defend the *patria* was higher than the feudal obligations of vassal to lord" (Kantorowicz 2012, p. 246). This new patriotism was nourished by values transferred from the *patria* of Heaven to the political communities of the earth. In the Crusades both knights and common soldiers could gain immediate entry into the Kingdom of Heaven, and in return for their sacrifice could obtain the crown of martyrdom in the other life. This is paradoxical, given that Christ sacrificed himself to prevent further sacrifices, but in seeking to put an end to the sacrificing of propitiatory victims, that sacrifice had the unintended consequence of becoming a form of sacrifice in which the core figure was the martyr, a pattern that was to continue into the idea of the national hero in modern times (Beriain 2011).

Fighting and dying for one's homeland came to be seen as a victory and remedy for the soul. Those who died in a campaign for their brothers (*sacrificio pro fratribus*), in a "holy war", emulated Christ's sacrifice for his brothers and sisters and offered themselves to God. Marcus Tullius Cicero put it this way:

> Who that is true would hesitate to give his life for her [one's native land] if by his death he could render her a service? (Cicero 2018, p. 49)

Those who fall on the field of battle for the *res publica* are glorified. What was good for Rome, the once capital of the world, served equally well for the incipient national monarchies in the kingdoms of Great Britain, France, Spain, etc. Loyalty to these new, territorial *patrias*, which included the subjects of the Crown, replaced the supranational bonds that had been used by the now fragmented Roman Empire. In this context, death for one's country as a "martyr" took the church as its reference point and adapted ecclesiastical formulae to the secular body politic. As pointed out by Kantorowicz, " ... death on the battlefield for the political *corpus mysticum* headed by a king who was a saint and therefore a champion of justice, became officially "martyrdom" (Kantorowicz 2012, p. 266). In this way, the *corpus mysticum patriae* was contrasted with the *corpus mysticum ecclesiae. Patria* was presented as an immortal, timeless entity invested with a unity that does not die, in the perpetuity of the people of the nation.

The tone of glory in self-sacrifice and the idea of transcending death have never been better expressed than in Pericles' funeral prayer for the Athenians who fell in the first year of the Peloponnesian War, composed by Thucydides:

> "They gave their lives, to her and to all of us,
> and for their own selves they won praises that never grow old,
> the most splendid sepulchres -nor the sepulchres
> in which their bodies are laid, but where their glory remains eternal in
> men's minds, always there in the right occasion to stir
> to speech or action. For famous men have the whole earth
> as their memorial: it is not only their inscriptions on their graves
> in their own country that mark them out; no, in foreign lands also,
> not in any visible form but in people's hearts, their memory abides
> and grows. It is for you to try to be like them. Make up your minds
> that happiness depends on being free, and freedom depends
> on being courageous. Let there be no relaxation in face of the perils of war". (Book II, Thucydides 1959, p. 121)

5.1.2. The nation as a New "Sacred Form" of Modernity with Its Altars, Monuments and Sacrificial Commemorations

The referents of the idea of "nation" are political power and "things national". It is a specific type of *pathos* which is linked to groups of humans united in a "community of people who share a common language, or religion, or common customs or political memories; such a state may already exist or it may be desired. The more power is emphasized, the closer appears to be the link" (Weber 1978, p. 327). However, among all the intangibles of the glue binding the community that this *pathos* establishes there is something missing; and that something is the sacrifice of a national hero who gives his/her life for the nation (Marvin and Ingle 1999, vol. 2, p. 63). Taking the later works of Durkheim and Girard as a reference point, Caroline Marvin points to the flag as a modern totem. The members of the totem group face, sooner or later, the limits of what is familiar and known and they reach the border, an area of confusion where identities are exchanged between *insiders* (nationals) and *outsiders* (immolated national heroes), and cross over. The crossing is violent and bloody, sacrificial in a word. *This dramatic encounter with death marks the exact border of the community.* The act of crossing establishes a clear contrast between who is inside and who is outside the community. Border crossers become *outsiders*, dead (immolated national heroes) to the community. The flag marks the point of their crossing. It is the symbol of those who have crossed, of devotees transformed. The community celebrates and reveres its insiders turned outsiders, taking steps lest they come back and punish those who did not cross over. From within the boundaries, the community fears and worships these outsiders (fallen heroes) it consumes to preserve its life (Marvin and Ingle 1999, p. 67). The community welcomes these returning border-crossers who have sacrificed themselves on the altar of the nation back to the fertile center by removing the mark of death they carry in piacular rituals in which there is a collective communion. Thus, "a hero's death for the freedom and honor of our people is a supreme achievement that will affect our children and children's children. There is no greater glory, no worthier end than to die this way. Additionally, to many, death gives a perfection that life would have denied them" (Weber 1995, p. 724).

In the narrative of the nation as told by nationalism there is a Golden Age, a driving myth which is projected as utopia into the future, as opposed to the historical contingencies of the present, but which is only attainable through commitment and self-sacrifice on the part of its members. This is what the nation continually defends, remembers and celebrates (Smith 2003, p. 218). The sentiments and symbolism expressed by Pericles served as models for shaping the political solidarity and civic nationalism of the Enlightenment and, by extension, of the French Revolution. The symbols inscribed and hinted at on monuments, the land and its occupants as an intangible sepulchre and the emulation of the courage of those who sacrificed themselves shaped a "secular religion" of the general will (Mosse 1975, vol. 2, pp. 71–72; 1990, pp. 33, 36, 38) in which the people of the nation venerate themselves. As stated by Bauman: "The hero's death was transcended, just as the death of the martyr had been—this time not by the salvation of the immortal soul of the dying but by the material immortality of the nation" (Bauman 2005, p. 44).

The modern reappropriation of sacrifice was to result in a new twist in the semantics of death. The existential nature of humanity described by Heidegger as "being for death" shifted towards "being for killing". But while dying is a solitary act, killing takes two. *Mortui viventes obligant*, so to justify this society introduces a new political performativity through a number of appropriating events that make up a national mystique, a community of worship in which great importance is given to the fallen within a politics of the masses, with new identifying totems and monuments—pyramids, obelisks, towers, statues, sarcophagi (Koselleck 2020, p. 66; Casquete 2020, p. 295 et seq.) that represent the eternity of time and in which the killed and the fallen are welcomed: heroes, victims, martyrs, conquerors, militants and, perhaps, the vanquished. This new culture of commemoration seeks to elevate human beings above their day-to-day routines by provoking feelings of fear, as if they were in a sacred temple linked to a community of worship. In pre-revolutionary times death was represented not as an ending but as a transition to another world, but it was presented as differentiated in terms of estates, i.e., kings, princes and warrior heroes were the bearers of that perpetuity that "never dies". However, with the onset of the Age of Modernity, from the French Revolution onwards, two major changes took place (Koselleck 2020, pp. 71–72): on the one hand the ideas of the intramundane representation of death came to carry more weight, i.e., a decline in the Christian interpretation of death left a gap for meanings based on purely political and social reasons. Post-Christian writings and linguistic forms refer to the earthly future of each state or people. On the other hand, earthly immortality, hitherto reserved for the great, became generalized in the name of all. This intramundane viewpoint was followed by a dismantling of the system of estates, i.e., a democratization of death. The attributes enshrined in St George as a helper, a savior and a monarch were superseded after World War I by the *Tomb of the Unknown Soldier* (Mosse 1990, pp. 36–38), who answers for the nation of which he forms part. There is an inward equality of the fallen for the homeland—a national homogenization in the face of death—but no outward equality. In a comment filled with acid irony the lead character in Charlie Chaplin's film *Monsieur Verdoux* (played by Chaplin himself), which was not too far removed from the arguments set out here, asserts that individual killers spark widespread rejection in the collective consciousness but mass killing sanctifies and creates heroes. In mass deaths there is a transfiguration of the meaning of death itself.

5.2. Landmarks in the Anti-Sactificial Narrative of the Sacralization of the Person

After the French Revolution, and above all after the two world wars and the horror of the Shoah, the legacy of the anti-sacrificial narrative was recovered in civil life, in dynamic tension with the sacrificial narrative of the nation. *This post-sacrificial narrative postulates that it is a crime to kill the victim (a human person) because it is sacred, so there has been a change from the sacrificial sacralization of the victim to the anti-sacrificial sacralization of the human person.* This is something that appears quite clearly in a publication by Emile Durkheim in 1898, brought up by Hans Joas (2017)[4], and has been confirmed institutionally in the various declarations of human rights (1776, 1789 and 1948). During the unrest that arose from the scandal of the Dreyfus Affair in 1898, Durkheim wrote:

> "This human person (*personne humaine*), the definition of which is like the touchstone which distinguishes good from evil, is considered sacred in the ritual sense of the word. It partakes of the transcendent majesty that churches of all time lend to their gods; it is conceived of as being invested with that mysterious property which creates a void about sacred things, which removes them from vulgar contacts and withdraws them from common circulation. And the respect which is given it comes precisely from this source. Whoever makes an attempt on a man's life, on a man's liberty, on a man's honor, inspires in us a feeling of horror analogous in every way to that which the believer experiences when he sees his idol profaned. Such an ethic is therefore not simply a hygienic discipline or a prudent economy of existence; it is a religion in which man is at once the worshipper and the god". (Durkheim [1898] 1973, p. 46)

This idea is not foreign to Christianity. Durkheim himself states this in his 1898 publication: "Christianity demonstrated in its inner faith, in the personal conviction of the individual, the essential condition of divinity ... The very center of moral life was thus transported from the external to the internal, and the individual was thus elevated to be sovereign judge of his own conduct, accountable only to himself and to his god" (Durkheim [1898] 1973, p. 52). It is an error to present this sacralization of the human person and its associated moral anchoring as antagonistic to Christian morality. Indeed, it derives precisely from that morality. By taking this on board we are not denying our past but continuing it.

Several of Jeffrey Alexander's more recent works (Alexander et al. 2004, 2013; Alexander 2013) contrast the two narratives mentioned here. On the one hand the tragic/apocalyptic narrative is based on aggressive heroism that vanquishes and kills enemy forces, so success is measured as the highest possible number of enemy casualties. This results in civilian and military victims who are irremediably traumatized by events that have caused them suffering, and thus turns trauma into an essential characteristic of their lives and circumstances. On the other hand, under the defensive/progressive narrative, soldiers risk and sometimes give their lives helping to retrieve wounded comrades. The significance of one's own suffering and that of others push them into performing acts of salvation, healing and care of victims and acts to improve the world. Mass deaths in war have been seen and morally justified as sacrifices *pro patria* under the tragic/ apocalyptic narrative. However, this changes "when narratives of triumph are challenged, when individual deaths seem

---

[4] Viviana Zelizer has studied the social processes that have led to the sacralisation of the child (Zelizer 1985). By the same author (Zelizer 2015), Part One: Valuation of Human Lives, pp. 35–123.

worthless or polluted, when those who have fallen are seen not as sacrificing for a noble cause but as wasted victims of irresponsible chicanery, that wars can become traumatic indeed" (Alexander 2013, p. 3), as occurred in the Vietnam war, which did not end in victory for the US. With this interpretational framework, recent research by Richard Lachmann (2016, pp. 323–58) shows a transformation in the culture of sacrifice through an analysis of the significance of the Medals of Honor awarded by the US Government to the families of those fallen in combat between 1861 and 2014. In the "pyramid of honor" that pays homage to awardees up to 1918 priority was given to the valor and intrepidness of soldiers in risking their lives beyond the call of duty. From 1963 onwards the post-Vietnam narrative changed the significance of the medal, emphasizing "noncombatant bravery", manifested through feats that resulted in the saving of military and civilian lives in fire-fights and bombings. The priority became avoiding victims, saving lives, rather than manufacturing victims. This was a major narrative shift in the cult of the nation and its commemorations. It also represents a way of paying military honours to soldiers in both victory and defeat, and ultimately undermines the *ethos* of self-sacrifice in the narrative of tragedy. However, in today's info-wars (*"from soldier to drone driver"*), combatants face each other in asymmetric conflicts in which one side is the *hunter*, armed with sophisticated weapons and technical systems, and the other is the *hunted*, equipped with inferior technology that leads it to assume a heroic role in the face of unseen death (Zulaika 2020).

Ullrich Bröckling (2020) detects two opposing tendencies in today's societies: on the one hand, from the 1980s onwards the attribute of being "post-heroic" has become increasingly plausible in various contexts, and on the other hand not a day goes by without new heroes and heroines appearing or old heroes being revived as emblematic figures or idealized human beings. Considering that we live in and project horizons of normative expectations and orders which are no longer hierarchical but heterarchical and therefore different from tradition, and more flexible frameworks of classification as conformity or deviation towards what is considered normal (Zerubavel 2018), the position of the individual in a highly complex, technified society, the models of leadership and, of course, the problem of the spirit of self-sacrifice and abnegation (and with it the attitude towards death) have changed drastically, so it is unlike the position of the heroes of the sagas, the redeeming heroes and the holy heroes described by Wilhelm Wundt ([1912] 1990, p. 335 et seq.). Western societies are no longer in a position to mobilize large numbers of people to give their lives "heroically" in the name of the nation's "tribal gods" (Isaacs 1975; Marvin and Ingle 1999). Modern-day heroes and heroines are characterized much more by their nonconformist positions critical of obedience. The impulse towards heroism manifests itself as civil courage, e.g., in Tiananmen Square in 1989, among fire-fighters and police officers on 9/11, in the activists currently fighting against climate change, against male dominance and in the *Black Lives Matter* movement and in the essential workers who have saved thousands of lives in the coronavirus crisis. At the same time heroism has become democratized and become an everyday phenomenon (*"just for one day"* as the David Bowie lyric puts it and as in Andy Warhol's statement that nowadays everyone can have their *"fifteen minutes of fame"*). This is related to changes in the ambit of creativity in modern societies, as creativity in post-modern times (unlike the tradition in which creativity was reserved for an elite) brings together two important, widespread issues: the subjective desire for creativity, i.e., the desire to be creative, and the objective imperative to be creative, i.e., the idea that one must be creative (Reckwitz 2012, p. 13).

6. Conclusions

The emergence of heroic and post-heroic figures begs the question of who needs such figures and why (Habermas 2003, p. 43). The answer may lie in different perceptions of crises and different desires for normalization; because wherever heroes appear they are a sign of underlying problems (Bröckling 2020). Heroes are symptoms of a social crisis, but that does not mean that they are the solution to it. Indeed, it is in the clash between heroic and post-heroic image guidelines that some of the lines of conflict of contemporary societies lie.

The genealogy of the various forms of sacrifice shows that as a communication link between the profane and the sacred which appeared with the emergence of farming societies, distinct from the idea of offering and with the immolation of a victim at its core, it has resulted in the creation of a powerful tragic/apocalyptic narrative whose influence can still be felt today in both religious and secular terms, within which the nation has created its own worship community.

Within the heterodox tendencies arising out of Hebrew monotheism that postulate the figure of a divine savior, there was strong criticism of immolation as a device and an affirmation of the ethical aspect of defending sacrificial victims. The influence of this criticism also spread over time in secular terms, giving rise to a no less important progressive/pacifist narrative. However, the dynamic tension between these two narratives leads to major lines of conflict in present-day societies.

Funding: This translation was funded by I-Communitas. Institute for Advanced Social Research (Public University of Navarra). This research was funded by the National Project "Variedades de la experiencia creativa y modelos de sociedad" (REF: CSO2017-85052-R) granted by Ministerio de Economía y Competitividad (Spain).

Conflicts of Interest: The author declares no conflict of interest.

References

Primary Sources and Collections

Biblia de Jerusalén. 1998. Bilbao: Desclée de Brouwer.

Secondary Sources

Akenson, Donald Harman. 1992. *God's Peoples. Covenant and Land in South Africa, Israel and Ulster.* Ithaca: Cornell University Press.

Alexander, Jeffrey C. 2013. *Trauma. A Social Theory.* Cambridge: Polity Press.

Alexander, Jeffrey C., Ron Eyerman, and Elizabeth Butler Breese, eds. 2013. *Narrating Trauma. On the Impact of Collective Suffering.* London: Palgrave.

Alexander, Jeffrey C., Ron Eyerman, Bernhard Giesen, Neil J. Smelser, and Piotr Sztomka. 2004. *Cultural Trauma and Collective Identity.* Berkeley: University of California Press.

Bauman, Zygmunt. 2005. *Liquid Life.* Londres: Polity Press.

Benveniste, Emile. 1969. *Le vocabulaire des institutions indo-européennes II.* París: Les Editions de Minuit.

Beriain, Josetxo. 2011. *El sujeto transgresor (y transgredido).* Barcelona: Anthropos-Siglo XXI.

Beriain, Josetxo. 2015. "Genealogía afirmativa" del hecho religioso en perspectiva sociológica. *Revista Española de Investigaciones Sociológicas, REIS* 151: 3–22.

Beriain, Josetxo. 2017. Las metamorfosis del don: Ofrenda, sacrificio, gracia, substituto técnico de Dios y vida regalada. *Revista Política y Sociedad* 54: 641–63.

Bröckling, Ullrich. 2020. *Postheroische Helden. Ein Zeitbild.* Francfort: Suhrkamp.

Butler, Judith. 2002. *Antigona's Claim. Kinship Between Life and Death.* New York: Columbia University Press.

Casquete, Jesús. 2020. *El culto a los mártires nazis. Alemania, 1920–1939.* Madrid: Alianza.

Cicero, Marcus Tullius. 2018. *Los deberes.* Madrid: Gredos.

Derrida, Jacques. 1995. *The Gift of Death.* Chicago: Chicago University Press.

Donald, Merlin. 1990. *The Origins of the Modern Mind. Three Stages in the Evolution of Culture and Cognition.* Cambridge: Harvard University Press.

Durkheim, Emile. 1973. Individualism and the Intellectuals. In *Emile Durkheim on Morality and Society.* Edited by Robert Neelly Bellah. Chicago: University of Chicago Press, pp. 43–57. First published 1898.

Durkheim, Emile. 1982. *Las Formas Elementales de la Vida Religiosa.* Madrid: Akal. First published 1912.

Eisenstadt, Shmuel Noah. 1986. *"Introduction" in The Origin and Diversity of Axial Civilizations.* Edited by Shmuel Noah Eisenstadt. Albany: State University of New York Press, pp. 1–29.

Erkens, Franz-Reiner. 2013. Herrschersakralität. In *Sakralität und Sakralisierung. Perspektiven des Heiligen.* Stuttgart: Editorial Franz Steiner.

Evans-Pritchard, Edward Evan. 1981. *La religión Nuer.* Madrid: Taurus. First published 1956.

Girard, René. 1982. *El misterio de nuestro mundo.* Salamanca: Sígueme (Abbreviated NMN). First published 1978.

Gordon, Neil. 1995. *Sacrifice of Isaac.* Nueva York: Bantam Books.

Habermas, Jürgen. 2003. Fundamentalism and Terror. A Dialogue with Jürgen Habermas. In *Philosophy in a Time of Terror. Dialogues with Jürgen Habermas and Jacques Derrida.* Edited by Giovanna Borradori. Chicago: The University of Chicago Press, pp. 25–43.

Hamayon, Roberte. 1990. *La chasse à l'âme. Esquissed'une Theorie Chamanienne Siberienne.* Nanterre: Societéd'Ethnologie.

Harrison, Jane Ellen. 1912. *Themis. The Social Origins of the Greek Religion.* Cambridge: Cambridge University Press.

Heidegger, Martin. 1984. *Gesamtausgabe.* Francfort: Vittorio Klosterman, vol. 53.

Hénaff, Marcel. 2010. *The Price of Truth. Gift, Money and Philosophy.* Stanford: Stanford University Press. First published 2002.

Henninger, Joseph. 1995. Sacrifice. In *The Encyclopaedia of Religion.* Edited by Mircea Eliade. London: Macmillan, vol. 11, pp. 544–57.

Hubert, Henry, and Marcel Mauss. 1981. *Sacrifice. Its Nature and Functions.* Chicago: Chicago University Press. First published 1898.

Irigaray, Luce. 1992. Belief Itself. In *Sexes and Genealogies.* Nueva York: Columbia University Press, pp. 23–53.

Isaacs, Harold R. 1975. *Idols of the Tribe. Group Identity and Political Change.* Cambridge: Harvard.

Joas, Hans. 2013. *The Sacredness of the Person. A New Genealogy of the Human Rights.* Washington, DC: Georgetown University Press.

Joas, Hans. 2017. *Die Macht des Heiligen.* Francfort: Suhrkamp.

Kantorowicz, Ernst H. 2012. *Los dos Cuerpos del rey. Un estudio de Teología Política Medieval.* Madrid: Akal.

Keenan, Dennis King. 2005. *The Question of Sacrifice.* Bloomington: Indiana University Press.

Kierkegaard, Sören. 1976. *Diario de un Seductor/Temor y Temblor.* Madrid: Guadarrama.

Koselleck, Reinhart. 2020. *Modernidad, Culto a la Muerte y Memoria Nacional.* Madrid: Centro de Estudios Políticos y Constitucionales.

Lacan, Jacques. 1992. *The Seminar of Jacques Lacan, Book VII. The Ethics of Psychoanalysis, 1959–1960.* Nueva York: Norton.

Lachmann, Richard. 2016. The culture of sacrifice in conscript and volunteer militaries: The U.S. Medal of Honor from the Civil War to Iraq, 1861–2014. *American Journal of Cultural Sociology* 4: 323–58. [CrossRef]

Leakey, Richard E. 1982. *Origins. What New Discoveries Reveal About the Emergence of Our Species and Its Possible Future.* New York: Dutton.

Leakey, Richard E. 1994. *The Origins of Humankind.* New York: Basic Books.

Levi-Straus, Claude. 1964. *El Pensamiento Salvaje.* México: Fondo de Cultura Económica.

Marvin, Caroline, and David W. Ingle. 1999. *Blood Sacrifice and Nation. Totem Rituals and the American Flags.* Cambridge: Cambridge University press.

Mauss, Marcel. 1971. Ensayo sobre los dones: Motivo y forma del cambio en las sociedades primitivas. In *Sociología y Antropología.* Madrid: Tecnos, pp. 155–267. First published 1924.

Mauss, Marcel. 1971. Mana. In *Sociología y Antropología.* Madrid: Tecnos, pp. 122–33.

Mosse, George Lachmann. 1975. *The Nationalization of the Masses.* New York: Howard Fertig.

Mosse, George Lachmann. 1990. *Fallen Soldiers.* Oxford: Oxford University Press.

Parsons, Talcott. 1978. The 'Gift of Life' and its Reciprocation. In *Action Theory and the Human Condition.* Nueva York: Free Press, pp. 264–99.

Pitt-Rivers, Julian. 1993. El lugar de la gracia en la antropología. In *Honor y Gracia.* Edited by Julian Pitt-Rivers and J. G. Peristiany. Madrid: Alianza, pp. 280–321.

Reckwitz, Andreas. 2012. *Die Erfindung der Kreativität. ZumProzessgesellschaftlicherÄsthetisierung.* Frankfurt: Suhrkamp.

Robertson-Smith, William. 1972. *The Religion of the Semites. The Fundamental Institutions.* Nueva York: Schocken. First published 1889.

Robertson-Smith, William. 1886. Sacrifice. In *Encyclopaedia Britannica.* Chicago: Encyclopædia Britannica, Inc., vol. 21, pp. 132–40.

Schwartz, Benjamin. 1975. The Age of Transcendence. In *Wisdom, Revelation and Doubt: Perspectives on the First Millennium B.C.* Cambridge: American Academy of Arts and Sciences, vol. 2, pp. 3–4.

Smith, Anthony D. 2003. *Chosen Peoples. Sacred Sources of National Identity.* Oxford: Oxford University Press.

Smith, Phillipe. 2005. *Why War? The Cultural Logic of Iraq, The Gulf War and Suez*. Chicago: Chicago University Press.
Stroumsa, Guy. 2009. *The End of Sacrifice*. Chicago: Chicago University Press.
Thucydides. 1959. *The History of the Peloponnesian War*. Hardmonsworth: Penguin.
van Baal, Jan. 1976. Offering, Sacrifice and Gift. *Numen* 23: 161–78. [CrossRef]
van Gennep, Arnold. 1986. *Los Ritos de paso*. Madrid: Taurus. First published 1909.
Weber, Marianne. 1995. *Max Weber. Una biografía*. Valencia: Alphons el Magnánim.
Weber, Max. 1978. *Economía y Sociedad*. México City: FCE.
Weber, Max. 1987. *Ensayos sobre Sociología de la Religion*. Madrid: Taurus, vol. 3.
Wundt, Wilhelm. 1990. *Elementos de Psicología de los Pueblos*. Barcelona: Alta Fulla. First published 1912.
Zelizer, Viviana. 1985. *Pricing the Priceless Child. The Changing Social Values of Children*. Nueva York: Basic Books.
Zelizer, Viviana. 2015. *Vidas Económicas. Cómo la Cultura da Forma a la Economía*. Madrid: CIS.
Zerubavel, Eviatar. 2018. *Taken for Granted. The Remarkable Power of the Unremarkable*. Princeton: Princeton University Press.
Zulaika, Joseba, and William Douglass. 1996. *Terror and Tabu. The Folies, Fables and Faces of Terrorism*. London: Routledge and Keegan Paul.
Zulaika, Joseba. 2020. *Hell Fire from Paradise Ranch. On the Front Lines of Drone Warfare*. Oakland: University of California Press.

Publisher's Note: MDPI stays neutral with regard to jurisdictional claims in published maps and institutional affiliations.

MDPI

Article

The Persistence of Sacrifice as Self-Sacrifice and Its Contemporary Embodiment in the 9/11 Rescuers and COVID-19 Healthcare Professionals

Javier Gil-Gimeno * and Celso Sánchez Capdequí *

I-Communitas. Institute for Advanced Social Research, Public University of Navarra, 31006 Pamplona, Spain
* Correspondence: fcojavier.gil@unavarra.es (J.G.-G.); celso.sanchez@unavarra.es (C.S.C.)

Abstract: The aim of this paper is to analyze the persistence of sacrifice as self-sacrifice in contemporary societies. In order to reach this goal, firstly, we discuss how in the Axial Age (800–200 B.C.E.) an understanding of sacrifice as ritual worship or a ritual practice that involves the immolation of a victim became less prevalent and a new understanding of sacrifice emerges. This new notion of sacrifice focuses on individual relinquishment and gift exchange, that is, on a person sacrificing or relinquishing him/herself as a gift that is given in an exchange relationship for protecting a greater good (a god, a community, a person, a nation, and so on). Secondly, we analyze how this new sacrifice formula had an important impact on the understanding of sacrifice. Most notably, it led people to conceptualize sacrifice as a project or as something that persons could intentionally embrace. Thirdly, and as a result of the previous processes, we attend to the secularization of sacrifice, not in the sense of a de-sacralization of this phenomenon but in the way of sacralization of the mundane realm and mundane things, such as intentional self-sacrificial acts, in social contexts where there is religious pluralism. Insight into how the notion of sacrifice is secularized is found throughout the classic works of Marcel Mauss and Georg Simmel, and these works are discussed in section three. Fourthly, we study the sacredness of the person as a clear type of secular religiosity that develops self-sacrificial forms. Two of these self-sacrificial forms are the actions of 9/11 rescuers and COVID-19 healthcare professionals. A short analysis of both will serve us to illustrate how self-sacrifice is embodied in contemporary societies.

Keywords: sacrifice; sacredness of the person; self-sacrifice; exchange; gift; relinquishment; secular religiosity

Citation: Gil-Gimeno, Javier, and Celso Sánchez Capdequí. 2021. The Persistence of Sacrifice as Self-Sacrifice and Its Contemporary Embodiment in the 9/11 Rescuers and COVID-19 Healthcare Professionals. *Religions* 12: 323. https://doi.org/10.3390/rel12050323

Academic Editor: Klaus Baumann

Received: 18 March 2021
Accepted: 29 April 2021
Published: 1 May 2021

> There is life when something is still also something else. There is death when something is only itself. A rigid tautology.
>
> **Roberto Calasso**

> Le sacrifice est l'homme.
>
> **Sylvain Lévi**

> [It] leads us after all to pose once more, in different forms, questions that are old but ever new.
>
> **Marcel Mauss**

1. Introduction

1.1. The Aim of the Paper

We start by pointing out that sacrifice has been a foundation in multiple and diverse forms in societies throughout history, and that sacrifice remains present in contemporary secular societies (Taylor 2007; Shilling and Mellor 2013). We adopt the premise that sacrifice can be considered an evolutionary universal. According to Paul Weiss. "Sacrifice occurs all the time" (Weiss 1949, p. 78). In this sense, sacrifice is studied here as a relevant experience of social life that is found in contemporary times. It is considered a meaningful part of human culture (Lévi 1966).

Our analysis is focused on one form of sacrifice: self-sacrifice. First, we study the emergence of self-sacrifice in Axial Age societies. Since that time, violent immolatory ritual sacrifice has been much less prevalent, while peaceful non-violent self-sacrifice has become more prevalent and has become one of the primary forms of sacrifice in contemporary times, especially in modern, secular, and individualized societies. We also show how the development of self-sacrifice led to an important change in the nature of sacrifice. Specifically, there was a clear transition from sacrifice as immolation to relinquishment and a gift exchange relationship. At the same time, there was a progressive secularization of the sacred and of the sacrifice, not in the sense of a de-sacralization but in the way of sacralization of the mundane realm and mundane things. As we will observe, the coexistence between otherworldly and innerwordly religiosities generates a scenario of "many altars", that is to say, of religious pluralism (Berger 2014). After analyzing this, we introduce the concept of the sacredness of the person understood as a secular religiosity that manifests itself in self-sacrificial forms. Two of these self-sacrificial forms are the actions of 9/11 rescuers and COVID-19 healthcare professionals. A short analysis of both will illustrate how self-sacrifice is embodied in contemporary societies.

1.2. The Persistence of Sacrifice

Before delving into an analysis of the waning of the idea of ritual sacrifice and the beginning of the conception of self-sacrifice, there is a preliminary issue that must be addressed. There are some authors, such as Agamben (1998) or Dworkin (2011) who reject the idea of the persistence of sacrifice in modern and secular societies. For Agamben: "In modernity the principle of the sacredness of life is thus completely emancipated, for sacrificial ideology, and in our culture the meaning of the term "sacred" continues the semantic history of *homo sacer* and not that of sacrifice" (1998, pp. 67–68). We agree with him when he states that the sacred remains present in contemporary societies, but we differ in respect to sacrifice. If we pay attention to the etymologic definition of our research's object, we can observe the term "sacrifice" comes from the Latin word *sacrificium*, which means "to make sacred". In this way, if the sacred sphere remains present in modern societies, we think there must be some formulas that allow the sacred to take social forms. In this sense, sacrifice would be one of these basic formulas that allow societies to activate the different processes and mechanisms of sacralization. Sociologically speaking, we think the sacred sphere needs sacrifice (in this case, self-sacrifice) for taking social form as a vehicle or mediation between the sacred and the not sacred spheres.

In discussing the persistence of sacrificing in modern advanced societies, we think it is important to note that social and human sciences have approached the analysis of the sacrifice in many ways. Some of them focus on the violence unleashed in sacrificial actions, for example, the studies by Girard[1] (1977, 2012), Derrida (1998) or Bloch (1992, 2015). Others, such as those of Tylor (2010), Milbank (1996) and Evans-Pritchard (1956) or Levi-Strauss (1966) emphasise the necessary presence of the gift and of the immolation. Other authors had been focused on different aspects: the key presence of patrilineal institutions for understanding the question of sacrifice (Jay 1992), their link with nationalism (Strenski 2002), their analysis such as a political problem (Tava 2018), or the sacrificial role played by national flag (Marvin and Ingle 1999), and so on. The diversity of voices that have analyzed (and remain to analyze it yet) the sacrifice and the variety of conclusions these works have reached lead us to think that our research object has survived to the different social metamorphosis and, for this reason, remains still present in contemporary social life. *We start with the premise that sacrifice can be considered an evolutionary universal which is involved in the different social scenarios, acquiring multiple and diverse forms.* According to Paul Weiss. "Sacrifice occurs all the time" (Weiss 1949, p. 78). In this sense, sacrifice is studied here as a relevant experience of social life, also in our contemporary social life. It is considered a meaningful part of the human culture (Lévi 1966).

We contend that Donald's methodological approach involving the principle of the conservation of gains is a pertinent way of achieving this paper's goals. In two of his major publications (Donald 1991, 2012), Donald presents a model of evolution based on the assertion that "previous adaptations are preserved" (Donald 2012, p. 54), in line with the "principle of conservation of gains" (Donald 2012, p. 54). It can therefore be deduced that subsequent stages of the process of evolution do not entail the disappearance of gains or the characteristic features of previous stages, but rather a reconfiguration of those actual features and their modes and possibilities of both action and accessing meaning. Through this theory, Donald seems to be saying that once a social phenomenon appears on the social scene, and true to Lavoisier's principle of the conservation of matter, it tends to transform rather than be destroyed. Thus, earlier gains are not only an essential part of the genetic code of subsequent ones, but they can also survive the transition from one stage to another and articulate themselves via different "masks" (Sánchez Capdequí 2004), maintaining their right to struggle for a voice and for social recognition. Perhaps one of the biggest mistakes made by evolutionary theorists—including such prominent names as Comte (2009) and Spencer (2004), and even a neo-evolutionist like Parsons (2005)—has been to consider that only hegemonic representations are present at each stage of evolution. "Top or governing representations were thus not the only cognitive–cultural representations circulating in the human matrix as evolution moved forward; they were the ones with most influence at that stage" (Donald 2012, p. 54). In terms of our study, Donald's "principle of conservation of gains" has the following three clear implications: 1. It enables us to establish a methodological grounding for analyzing the persistence of sacrifice in current societies. 2. New sacrificial forms do not mean the death of previous sacrificial forms. Ritual or pre-axial sacrifices do not disappear in modernity. 3. Self-sacrifice (the sacrificial form we consider here) is not the only one present in contemporary societies, but one that coexists with other forms.

If this is so, why is it so difficult to recognize the persistence of sacrifice in current societies? We want to briefly analyze two causes: First, because around sacrifice—as around any social fact—there arise *pars pro toto* dynamics that restrict the social fact to its hegemonic or elementary features. In this case, we can observe a turmoil between ritual sacrifice (*pars*) and sacrifice (*toto*). According to this logic of action, there would be no place for sacrificial demonstrations (as self-sacrifice) beyond those developed during what Hénaff (2010) and Hamayon (1990) have called "The Age of Sacrifice" (i.e., within the context of pre-axial religiosity, a period that dates back to the transition from hunter–gatherer to agropastoral societies, from a nomadic to a settled lifestyle). It is therefore closely linked to the idea of a social life centered on crops and livestock husbandry. As noted by Hamayon (1990, p. 737), a second change takes place in such societies; this time in matters of religion or relationships with the sacred: the supernatural becomes vertical, introducing a world of hierarchies and relationships based on dependence (of the human world on the supernatural) and on filiation, which replaces the earlier notion of an alliance with the supernatural: "[w]e are truly entering a world of hierarchy and debt, which is precisely the world of sacrifice" (Hénaff 2010, p. 171). In these circumstances, ritual sacrifice acts as a mechanism for intermediation in that now-hierarchical scenario, where there is a chasm (Eisenstadt 1986) between the human and the divine. Sacrifice serves to open up communication between the two worlds: the mundane (represented by the sacrificer) and the divine (represented by the sacred object). This argument is consistent with the core program established by Marcel Mauss and Henri Hubert in their early work *Essai sur la nature et fonction du sacrifice* (1964), first published in 1898. Thus, from the perspective that equates sacrifice with "Age of Sacrifice", the end of public sacrifices (Stroumsa 2009) or the end of ritual sacrifices would mean the end of the social fact of sacrifice. However, we argue that the end of ritual sacrifice actually means its decline as a hegemonic representation or "total social fact", but not its disappearance as a phenomenon on the path taken by societies. Sacrifice has slipped from the dominant position that it held during its "Age", but that does not mean that it has disappeared from our global society. Accordingly, self-sacrifice is a post-ritual formula that emerges during the axial age and remains present in contemporary societies, coexisting with other sacrificial formulas—including ritual sacrifice.

Secondly, there is an analogy between the "process of civilization"—Elias ([1939] 1968)—and the premise of the end of sacrifice when we seek to explain the reluctance to recognize the persistence of sacrifice in modernity. According to this assumption, sacrifice would be a clear example of the stage of barbarity, and it has no place in secular, modern and civilized societies. This is the perspective supported—among others—by Agamben (1998) and Dworkin (2011). As suggested by Terry Eagleton: "That orthodoxy has been well-nigh unanimous in repudiating the concept of sacrifice as barbarous and benighted" (Eagleton 2018, p. 2). We contend that these views on sacrifice that understand it to be a barbarian practice commit a double mistake. First, they accept that there is no place for barbarity or violence in civilized societies. Scholars such as Adorno and Horkheimer (2002) and Bauman (1991), provide us with clear examples of the cohabitation between civilization and barbarity in modern—and theoretically civilized—societies. Secondly, as we assert here, sacrifice is a social fact. From this sociological perspective, what really affects any social fact—with sacrifice being no exception—is metamorphosis. Sacrifice mutates and adjusts to the different social and cultural realities in which it develops.

When analyzing sacrifice, scholars such as McClymond (2008) have posited that it requires a death or, at least, a victim. This question features in the classic discussions about our subject. For example, Mauss and Hubert (1964) share McClymond's approach. Other scholars, such as Robertson Smith (1886, 1972), consider a sacrifice to involve ritual offerings in which there is no immolation of a victim. Regardless of these discussions, it is important to say that the bulk of modern studies on sacrifice (Detienne and Vernant 1989; Cassirer 1955; Weiss 1949; Van Ackeren and Archer 2018; Herrenschmidt 1989; Tessman 2018) focus more specifically on the relationships of exchange *between the gift and the relinquishment (i.e., between what is given and what is taken)*. This means that "immolation" is not indispensable in this kind of analysis, or it does not perform the core role formerly played. At the same time, as already observed, this does not mean that violence or immolation have disappeared from sacrifice. *Narratives on sacrifice centered on violence- and gift-relinquishment share the same span of time within our societies, albeit in dynamic tension*. In fact, the paper's final sections briefly analyze two sacrificial (self-sacrificial) formulas in which the two aspects, namely, immolation of a victim and the relationship of exchange between gift and relinquishment, are crucial for understanding the sacrificial dimension whereby certain people or collectivities decide to *offer* their lives—or are ready to *offer* their lives—to save the lives of others in contexts of social stress or necessity. Having said that, in the two examples that we will briefly be analyzing, immolation is not the action axis, but the self-offering in terms of gift and relinquishment. This happens within a *process* of *sacralization of the person* (Durkheim 1973; Joas 2013), in which the human being simultaneously performs the roles of believer and god. In so doing, sacrifice breaks free from the traditional tendency toward *sacrificing to the god*, and prompts a new inclination focused on the *God's sacrifice* (and, more recently, personal self-sacrifice). Individuals surrender their lives in a radical act of personal sacrifice.

What, then, are the mainstays of the self-sacrificial formula to be analyzed here? We shall be dedicating a large part of this paper to answering this question. We should like to begin by identifying these three mainstays: firstly, *the ongoing acquisition of strength and prominence of the self in sacrificial practices*. This is directly linked to a decreasing precision in differentiating the classic roles of ritual sacrifice (i.e., victim, sacrifier, and sacred object); secondly, there is a *transition from a sacrifice focused on a victim's immolation and the hierarchical relationship between inner-worldly and otherworldly realms to a less hierarchical relationship based on the idea of exchange and focused on the question of gift and relinquishment*. Thirdly, *there is a secularization of sacrifice*, not in the sense of a de-sacralization, but rather as a metamorphosis in the sundry logics of sacralization that multiply themselves, taking different shapes that are not necessarily Historic (Bellah 1969) or Axial (Eisenstadt 1986; Bellah 2011), as they may also be Secular (Aron 1944). In this scenario, the inner-worldly realm becomes appropriate to sacralization. The sanctity of the person is one of these secular religiosities that fuels self-sacrificial formulas. We will illustrate this theoretical analysis of self-sacrifice and how it is embodied in contemporary societies by briefly analyzing this type of religiosity and two self-sacrificial forms linked to it: the actions of 9/11 rescuers and COVID-19 healthcare professionals.

2. Axial Religiosity, the Waning of Ritual Sacrifice and the Beginning of Self-Sacrifice

2.1. The "End" of Sacrifice? Self-Sacrifice and the Metamorphoses of Sacrifice in the Axial Age

The Axial Age (Jaspers 2011) involved a great transformation in social and religious spheres between 800 and 200 BC. According to Shmuel N. Eisenstadt: "This revolutionary process took place in several major civilizations including Ancient Israel, Ancient Greece, Early Christianity, Zoroastrian Iran, early Imperial China, and the Hindu and Buddhist civilizations. Although beyond the Axial Age proper, it also took place in Islam. These conceptions were developed and articulated by a relatively new social element. A new type of intellectual elite became aware of the necessity to actively construct the world according to some transcendental vision. The successful institutionalization of such conceptions and vision gave rise to extensive re-ordering of the internal contours of societies as well as their internal relations" (Eisenstadt 1986, p. 1). Examples of this new intellectual elite are the Jewish prophets, Greek philosophers, Chinese literati or Hindu brahmins. These intellectuals attempt to re-order the world in terms of the model of the "ideal man" (Eisenstadt 1986, p. 5). *This change implies a new relationship between individual and the world. The individual starts to be in the focus of social concern*. Insofar as pertains to the purposes of our study, the Axial Age marked a transition from religious acts focused on worship (basically sacrifice) to acts focused on *logos,* the word of god as revealed to a chosen few. Thus, axial-age religiosity focused on the word. This "linguistification of the sacred" (Habermas 1984) marking a change in the way in which believers related to their gods. In this scenario, the (holy) book gradually supersedes sacrifice as a privileged or hegemonic way of accessing or communicating with the other world (Assmann 2009; Gil-Gimeno 2021). Intermediation with the sacred no longer involved preferably the immolation of a victim but throughout the truth of a word revealed. *No longer involved preferably ritual but introspection*. Despite this metamorphosis occurs, it would be wrong to think that the end of the Age of Sacrifice or the ritual sacrifice meant the end of sacrifice as a social phenomenon. The "principle of conservation of gains", as discussed by Donald (2012), must be taken into account. This is worth repeating: we do not believe that the decline in sacrifice as a hegemonic manifestation of religion meant the end of sacrifice as a social fact. Stroumsa (2017) writes of the "end of public sacrifices", not the end of sacrifice *per se*. Therefore, we are facing, at the same time, a *metamorphosis* of sacrifice and toward a waning of the ideal of ritual sacrifice.

2.2. The Emergence of Self-Sacrifice. The Example of the Death of Jesus of Nazareth

A clear example—not the only one—of this axial transition in terms of sacrifice is the death of Jesus of Nazareth on the cross. It can be seen as the symbolic culmination of the "Age of Sacrifice" (Keenan 2005; Girard 2012; Theissen 1999), understanding this death as the sacrifice of God himself, this is, God the Son giving up his life to God the Father: "Father, into Thy hands I commend my spirit" (Luke 23–46). This viewpoint presents Jesus solely in his divine dimension. Thus, his death on the cross can only be seen as a sacrifice, but one which, paradoxically, seeks to blow away the very logic of ritual sacrifice and turn it into a pagan act which is therefore not fit for forging communication links with the supernatural: it is the ultimate sacrifice, or the sacrifice of sacrifice itself (Nancy 2003; Keenan 2005). However, this approach raises doubts as regards the process of consecration, because if the victim is already consecrated, he cannot be consecrated through the sacrifice. Similarly, the action of the sacrifier in offering up the sacred thing is closer to sacrilege than sacrifice.

This identification of Jesus of Nazareth solely with his divine side triumphed when Christianity became the official religion of the Roman Empire (with the Edict of Thessalonica, issued by Emperor Theodosius in 380 AD), subsequently becoming institutionalised and acquiring hegemony in the Western societies in the Middle Ages. It was symbolised magnificently in the expression *Sacrificiorum aboleatur insania* ("let the madness of sacrifice end!"), which appears in the Codex Theodosianus (438 AD) (Belayche 2001). This expression also marks the shift from a sacrificial religiosity to an axial one, based on the word and focused on the self, and the willingness to make a gift to another and, at the same time, to relinquish their life for getting a greater good. However, this identification reveals certain trouble, as pointed out by Sophie-Grace Chappell:

"What happened on the Cross is a sacrifice in something very like the way that the sacrifices laid down in Leviticus are; only more so. The point of a sacrifice, in Leviticus (see especially Chapters 1–7), is to take away sin, and recreate, by establishing (in blood) a new contract or "covenant", the human–divine relationship that has been disrupted. The more perfect the priest, and the more perfect the victim, the truer and the more effectual the sacrifice. Yet no earthly victim, and no earthly priest, can really be *completely* perfect; only God himself can be that. The ultimate sacrifice, then, the sacrifice that finally and definitively "fulfils the Law" (Mt 5.17), must be one *like* the Levitical sacrifices, and yet of an entirely different order, offered by a priest who, like the mysterious Melchizedek (Ps 110.4, Hebs 6–7), is no Levite at all, and seems to have no clear human origin that anyone can trace. Since literal perfection is required both in priest and in victim, and since only God is literally perfect, this ultimate sacrifice must be *God offering God*; but since the sacrifice has to be one in which the priest represents humanity before and to God, it must also be *man-God offering man-God.* And that, says Hebrews, *is precisely what Jesus does on the Cross.*" (Chappell 2018, p. 18)

The fact that this interpretation of the divine nature of Jesus gradually became hegemonic does not mean that there were no other interpretations that emphasized his human side in early Christian communities[2]. Indeed, this latter interpretation can even be justified also by reference to the canonical texts of the Church, e.g., the Gospel of John (John 1: 14), where Jesus is referred to as follows: "And the Word was made flesh and dwelt among us"[3]. He is thus seen as God become man and not exclusively as the son of God. This dual nature (divine and human) gathered into one person broaden the range of interpretations in regard to his death and its sacrificial implications, taking it beyond the logic of the sacrifice of sacrifice (Keenan 2005) or the sacrifice of God (Nietzsche 1967), because it could be argued that it is not the God who gives up his life but the man (or the God-as-man), i.e., *that it is a person who empties himself, who offers himself up completely, who relinquishes his life* in an exercise of *kénosis*.

From this viewpoint, Jesus is not the sacred thing (sacred being) but rather a propitiatory victim who offers himself or is offered up[4] in a self-sacrificing process for a greater good, i.e., the forgiveness of original sin. By this action, he either transcends his human dimension (returning to the divine) or becomes consecrated (thus becoming a sacred being through his act of sacrifice). Under both views, his action can be seen as a sacrifice in all lights. As mentioned above, this interpretation of the nature of Jesus was popular among some early Christian communities. Specifically, those considered as martyrs[5] acted on the basis of the idea of *Imitatio Christo*, seen as a way of consecrating themselves and accessing the sacred. From this, it can be deduced that these *early Christians saw Jesus as a man, albeit a great man, who attained transcendence by selflessly giving up his life.* The *Imitatio* of Jesus as God could be seen as sacrilege, but the *Imitatio* of Jesus as a man opened up the way to a process of consecration throughout an exercise of relinquishment and gift. Thus, it can be argued that what these martyrs did was not sacrilege (i.e., acting as if they were gods) but quite the opposite, i.e., consecrating themselves through their actions as their spiritual leader had taught them. This would make them victims who self-immolate and relinquish their lives for the salvation of the world, i.e., for the collective (sacrifice). In this sense, Stroumsa pointed out: "Elsner analyzes the mosaics of Santa Maria Maggiore in

Rome dating from Pope Sixtus III, in the 430 s, and the sacrificial processions in the mosaics of Sant´Apollinaris Nuovo. According to him, these mosaics demonstrate not only the symbolic polysemy of Christian sacrifice, but also the radical abolition of the ideology of reciprocity, even if the pagan gesture of sacrifice was still known. The martyrs and virgins no longer bring the sacrifice—they *are* the sacrifice. In effect, Christianity offers to every man and woman the possibility of becoming the sacrifice" (Stroumsa 2009, pp. 76–77). In fact, as it was pointed out by Rives (1995), human sacrifice was jointly considered by pagans and Christians a boundary between humanity and barbarism.

It is important to say that the archetypical and exemplary action of self-sacrifice developed by Jesus of Nazareth also provoked the emergence of the martyr that will replace the scapegoat in monotheistic religions, producing a metamorphosis in sacrificial narrative and oriented the sacrifice in a new, more spiritual (introspective), individualized and mundane way that it will be gradually developed and that it will reach their flashpoint in modernity.

Likewise, the death of Jesus of Nazareth on the cross is a clear example—not the only one, of course—of the metamorphosis that experiments with sacrifice in the Axial Age. In this sense, *we can observe a first transition from ritual sacrifice to self-sacrifice.* At the same time, sacrificial actions were marked by a number of changes in the own dynamics of sacrifice. First of all, as mentioned by Stroumsa (2017), sacrifices ceased to be openly public in nature and became more internal and spiritual (Cassirer 1955; Detienne and Vernant 1989; Keenan 2005; Duyndam et al. 2016) or introverted (Adorno and Horkheimer 2002). Sacrifice shifted away from the altar (Eagleton 2018) and became lodged more and more within each individual. The same occurred with religion and religiosity. Secondly, there was a change in the relationship between the sacrifier and the sacrificial victim; on the one hand, the sacrifier no longer played such an active role in consecration (though he/it continued to receive the benefits associated with the immolation of the victim). On the other hand, the victim gradually acquired more of a voice and greater autonomy, and thus began to be a more active party in the process of consecration. As a result, victims also began to receive the benefits of the sacrifice. In both the cases analyzed here, the victim no longer needs to be led to the altar by the sacrifier but is increasingly willing to go on his/her own initiative. Here we can start to observe a clear tension between what Ulrich Bröckling calls "chosen self-sacrifice" and "not chosen by oneself victim" (Bröckling 2020, pp. 230–31). The independence thus acquired is linked to the third and most significant difference with ritual sacrifice: *victims are no longer domesticated animals but persons who offer up their life willingly to save others* (Petropoulou 2008). In this sense, Schiller (1845) or Ingolf Dalferth speak about altruistic sacrifice for defining this process as "[s]omething done for others, which leads to the loss of one's own life". (Dalferth 2010, p. 83). In the fourth section, we will analyze how the axial turn toward self-sacrifice (focused on the field of historic religions) becomes the "highest form of sacrifice" (Weiss 1949, p. 80) in modern and secular societies.

Summarizing, in the Axial Age we attend the emergence of self-sacrifice and to a transition from ritual sacrifice to this new sacrificial formula (self-sacrifice): "A key moment in the evolution of sacrifice arrives when the victim themselves becomes conscious of their condition, and in doing so assumes agency of the event (. . .) What was a process to be endured becomes a project to be executed" (Eagleton 2018, pp. 50–51). This is the starting point of a clear metamorphosis in sacrifice that, step by step, becomes more secularized and focused on gift and relinquishment social fact.

3. Gift, Relinquishment and Exchange

3.1. Secularization, Exchange and Sacrifice

In their *Essai sur la nature et fonction du sacrifice* (1898), Hubert and Mauss established a clear connection between the sacred and profane, substantiated in the victim's immolation. Nevertheless, in a work authored alone and titled *Essai sur le don* (2002), first published in Année Sociologique in 1924, Mauss adopted a new methodological bias to explain the connection between the two realms. Whereas the first publication is essential for a sociological systematization of ritual sacrifice, and therefore for understanding its outlines and consequences, the second one has certain key elements that allow us to analyze, on the one hand, the persistence of sacrifice over time and, on the other hand, the different metamorphoses that this social fact has undergone. We therefore contend that the concept of gift plays an important role. We want to discuss *The Philosophy of Money* (2004) by Georg Simmel, where he develops a secularized notion of sacrifice based on the premise of relinquishment.

Why do the concepts of gift and relinquishment in Mauss and Simmel allow us to better understand the self-sacrificial forms? Through them, the collective dynamic of giving and receiving—institutionalized by social constructs like potlach in ancient societies or by the monetary market in advanced societies—becomes a key axis of sacrificial practice. In the previous section, we analyzed the emergence of self-sacrifice and how the arrival of axial religiosity leads to a transition from ritual sacrifice to self-sacrifice. Our aim in this section is to study another double process linked to the metamorphosis of sacrifice: the secularization of the phenomena and the acquisition of relevance by the gift and the elements of relinquishment at the expense of those linked to immolation.

Before analyzing the contributions made by Mauss and Simmel, we need to make two comments: the first one is that both contributions are aimed at social institutions that have not been considered as religious in a historical sense (Bellah 1969), like the market or the primitive potlach. The second one is that the two proposals focus on the idea of exchange as a common denominator. Let us now briefly discuss them both.

Firstly, the secularization of sacrifice does not mean a decline in belief or religious practice (Casanova 2001), but a reorientation of religiosity towards the profane or secular realms. For Durkheim: "[s]acred things are not simply those personal beings that are called gods or spirits. A rock, a tree, a spring, a pebble, a piece of wood, a house, in a word anything, can be sacred [. . .]. The circle of sacred objects cannot be fixed once and for all; its scope can vary infinitely from one religion to another" (Durkheim 1995, pp. 34–35). Underlying these words is an idea that can also be found in the studies by Robertson Smith (1972) and Mauss and Hubert (1964): something is rendered holy by the collective sentiment with which it is addressed. Through the development of secular religiosities—as illustrated through the sanctity of the person—sacrifice has not only transcended the ritual dimension, but has also superseded traditional religiosity (Bellah 1969). This happens because sacralization processes have also undergone changes in their historical course. "Classic" religious formulas coexist in modern societies with other secular religiosities (Aron 1944). Some examples of these are the sanctity of the person (Durkheim 1973; Joas 2013), civil religion (Bellah 1967; Giner 1993), revolutionary cults (Mathiez 2012), and public religions (Casanova 1994).

Secondly, the narrative of the transition from barbarity to civilization has several implications for the sacrifice. We should not confuse barbarity with violence, or civilization with the absence of violence. Examples such as 9/11, the Holocaust or the Gulag are proof of this. It is true that we live in civilized societies or in societies in which a "process of civilization" has taken place, but this does not mean we live in non-violent societies. In the transition from barbarity to civilization, societies adopt mechanisms for controlling violence. The main one has been state bureaucratic rationalization (Weber 2004), together with the subjective rationalization of the affective household (Elias [1939] 1968). Yet this does not mean that other societies in the past have not created their own mechanisms for achieving the same goal. This is the position defended by Girard (1977, 2012), whereby ritual sacrifice is one of the first social attempts to harness pure violence through ritual violence. Ritual sacrifice would therefore be one of the first milestones in the civilization process, and not a clear symbol of barbarism.

We will now focus on the contributions made by Mauss ([1924] 2002) and Simmel (2004). With respect to the former, the discussion addresses two issues: first, the collective dimension involved in the gift institution, and second, the spiritual dimension involved in the object of exchange.

3.2. Le Don, Mauss and Sacrifice

Mauss stated the following in *The Gift* (Mauss [1924] 2002): "First, it is not individuals but collectivities that impose obligations of exchange and contract upon each other. The contracting parties are legal entities: clans, tribes, and families who confront and oppose one another either in groups who meet face to face in one spot, or through their chiefs, or in both these ways at once. Moreover, what they exchange is not solely property and wealth, movable and immovable goods, and things economically useful. In particular, such exchanges are acts of politeness: banquets, rituals, military services, women, children, dances, festivals, and fairs, in which economic transaction is only one element, and in which the passing on of wealth is only one feature of a much more general and enduring contract" (Mauss [1924] 2002, pp. 6–7). These words reveal the collective dimension of the institution of giving, while at the same time we can start to perceive its related moral or spiritual dimension.

The last question specifically arises when Mauss defines the gift exchange as a spiritual mechanism that obligates the "person to reciprocate the present that has been received" (Mauss [1924] 2002, p. 9); in other words, the exchange based on the gift, its essence, lies in the spiritual (symbolic) value present either in the objects or in the actual exchange process. "The thing received is not inactive (. . .) because the thing itself possesses a soul, is of the soul" (Mauss [1924] 2002, pp. 15–16). "In this system of ideas one clearly and logically realizes that one must give back to another person what is really part and parcel of his nature and substance, because to accept something from somebody is to accept some part of his spiritual essence, of his soul" (Mauss [1924] 2002, pp. 15–16).

Mauss ([1924] 2002) thereby implicitly expands the sacrifice's sphere of action when indicating that one of the main links in human and social experience is the obligation of giving and receiving. Ritual sacrifice would therefore be a precise formula adopted by this social "leading wire". In this shift, Mauss seems to retrace his 1898 steps by reporting that the gift institution is a more important element in sacrificial ritual practice than the immolation of the victim as a way of communicating between the mundane and supramundane. Likewise, the narrative shift Mauss takes allows us to refer to emancipation of sacrifice from its ritual manifestations of immolation. Exchange relationships still remain following the transition from ritual to self-sacrifice, when axial societies give way to secular religiosity. In fact, what in axial religiosity is perceived as ongoing proximity between divine and human realms, in modern societies becomes secular religiosity, which means correlation (Cassirer 1955) rather than a chasm, between sacred and profane realms.

The argument we uphold is the following: The ceremonial exchange relationships based on gift and relinquishment (and its full secular versions developed by Simmel) are a more appropriate cue for understanding the metamorphosis of sacrifice, and above all their persistence in secular and modern societies, than other forms related to the victim's immolation for filling the vacuum between the mundane and supramundane realms. This does not mean, for example, that the element of immolation disappears, because sacrificial practice (as in the examples to be analyzed later) could sometimes require the giving of life for keeping alive the circle of sacred things. This idea is indeed closely linked to what occurs in exchange mechanisms, as we shall see in the work by Simmel (2004).

It needs to be recognized that this relevance of exchange and reciprocity in sacrificial ritual practice has already featured in the analyses made by prominent scholars such as Evans-Pritchard (1956); Tylor (2010); Van Baal (1976); and Robertson Smith (1886, 1972). The last of these focuses on a governing principle of social life: "those who sit at meat together are united for all social effects" (Robertson Smith 1972, p. 269). The idea here is that sacrifice makes divine beings and mortals into commensals. Robertson Smith holds that this is the main function of sacrifice, whereby it is linked not so much to consecration as to sociability around a table (commensality). This viewpoint is also developed by Seaford (2004) in the context of Greek Sacrifice: "Though always dedicated to a god [sacrifice], it was a communal event, with the meat shared among the participants and with only the bones and the fat burned for the god, as against the more normal Mesopotamian case where the sacrifice was dedicated primarily to the god and only the king or priests could partake." (Bellah 2011, p. 669).

3.3. Economic Exchange, Simmel and Sacrifice

Although sacrifice is not the core element in the *Essai sur le don*, the analysis conducted by Mauss provides us with interpretative elements for better understanding this social fact, its metamorphosis, and its persistence in modern and secular societies. Simmel's *Philosophy of Money* (Simmel 2004) also offers us new interpretative elements, but on this occasion, sacrifice becomes a key concept for articulating the economic exchange theory.

For Simmel: "The object acquires its practical value not only by being in demand itself but through the demand for another object. Value is determined not by the relation to the demanding subject, but by the fact that this relation depends on the cost of a sacrifice which, for the other party, appears as a value to be enjoyed while the object itself appears as a sacrifice" (Simmel 2004, p. 77). Sacrifice, understood as relinquishing something valuable—a thing, person, symbol, and so on—for achieving something, becomes—for Simmel—not only the measurement of something's value, but also the measurement of the exchange value itself (in this case monetary exchange). *Simmel introduces us to sacrifice completely devoid of the element of "immolation", focused on exchange relationships and, above all, the act of relinquishment* (the meaning of sacrifice in its own terms) *that it always requires.* People relinquish the thing sacrificed, which is given as a gift for achieving something. *Simmel's proposal has two clear ramifications for our work: firstly, it awards a moral, spiritual, and even religious dimension to economic exchange through sacrifice; secondly, it certifies the existence of a secularized version of sacrifice based on the notion of relinquishment.* In so doing, Simmel agrees with another modern analysis centered on sacrifice that emphasizes the importance of relinquishment. This is a highlight in the works of Herrenschmidt (1989), who says that what is relinquished is offered to another, Bataille (1992), who focuses on the relationship between relinquishment and giving in the sacrificial practice, and Marcel Detienne and Vernant (1989), who states the following: "[w]ith the appearance of sacrifice everything changes, for the most obscure or even the crudest of sacrificial acts implies something unprecedented: a movement of self-abandon" (Detienne and Vernant 1989, p. 20). We can find traces of our argument in the works of Dalferth (2010); Schiller (1845); and Cassirer (1955).

For Simmel, sacrifice plays a significant part in the economic exchange equation, whereby exchange would be worthless without sacrifice. "But here we overlook that sacrifice is by no means always an external obstacle but is the inner condition of the goal itself and the road by which it may be reached (. . .) In brief, the inhibiting counter-motion, to eliminate which a sacrifice is required, is often, perhaps even always, the positive precondition of the goal. The sacrifice does not in the least belong in the category of what ought not to be, as superficiality and avarice would have us believe. Sacrifice is not only the condition of specific values, but the condition of value as such; with reference to economic behaviour, which concerns us here, it is not only the price to be paid for particular established values, but the price through which alone values can be established" (Simmel 2004, p. 82). In short, the works by Mauss ([1924] 2002) and Simmel (2004) show us how, through its development, sacrifice undergoes a secularization process focused on relinquishment and the exchange of gifts. The two dynamics analyzed in the previous sections (i.e., individualization and secularization) originate from the emergence of secular patterns and examples of self-sacrifice. The next sections analyze the sanctity of the person as a secular religiosity that is governed by new self-sacrificial formulas. We briefly analyze two: the actions undertaken by 9/11 rescuers and by COVID-19 healthcare professionals.

4. Self-Sacrifice, Modernity and Secularization Process

In the previous section, we said that we must avoid the easy identification of secularization with desacralization. Rather, what happens is that secularization inserts changes in the way human societies accomplish the task of making things sacred. With respect to our goal, we want to highlight two of these changes: firstly, the inner-wordly realm and mundane things become potentially goods for sacralization. This does not mean that innerwordly realm replaces the other-worldly realm as object of sacralization, but rather that these secularized religious forms and another traditional or "historic" (Bellah 1969) live together in modernity. Secularisation did not spell the end of religiosity but rather a crisis in what Bellah calls "historic religions" (Bellah 1969, p. 78) on the one hand and the rise of a new way of making things sacred on the other, characterised by "the collapse of the dualism that was so crucial to all the historic religions" (Bellah 1969, p. 79) and by the sacralisation of secular, mundane aspects. Thus, the scenario in the modern era can be defined as one of religious pluralism (Berger 2014) where the religious domain does not however hold hegemony over the life of society as a whole. Secondly, in the introduction, we spoke on the idea of "making sacred" one thing, object, being, and so on. Well then, the secularization processes influence both fields, the sacred as well as the profane. Self-sacrifice shapes a new constellation of meaning in which the bloody, immolatory and ritualistic forms are reshaped by non-violent, secular, forms of self-sacrifice.

Having said that, in this section, we want to analyze two specific dimensions of this new type of sacrifice. On the one hand, we study the sacredness of the person as one of the most important types of secular religiosity that develops self-sacrificial forms. On the other hand, we briefly analyze—as an example—two of these self-sacrificial forms: the actions developed by 9/11 rescuers and COVID-19 healthcare professionals. A short analysis of both will serve us for illustrating how self-sacrifice embodies in contemporary societies, this is, how sacrifice persists in contemporary societies throughout self-sacrifice.

4.1. The Sacredness of the Person

The study of the sacredness of the person can first be glimpsed in the mid 19th century in Ludwig Feuerbach's *The Essence of Christianity* (Feuerbach 1832), but it was Durkheim who first tackled the issue head-on in "Individualism and the Intellectuals" (Durkheim 1973), first published in 1898. He states that in modern societies "it is humanity which is worthy of respect and sacred" (Durkheim 1973, p. 48). However, it is important to point out that what is sacralised is humanity in the abstract sense (in its altruistic, universal dimension) and not a particular individual. In short, "this human person (*personne humaine*), the definition of which is like the touchstone which distinguishes good from evil, is considered sacred in the ritual sense of the word. It partakes of the transcendent majesty that churches of all time lend to their gods; it is conceived of as being invested with that mysterious property which created a void about sacred things, which removes them from vulgar contacts and withdraws them from common circulation. And the respect which is given it comes precisely from this source. Whoever makes an attempt on a man's life, on a man's liberty, on a man's honor, inspires in us a feeling of horror analogous in every way to that which the believer experiences when he sees his idol profaned. Such an ethic is therefore not simply a hygienic discipline or a prudent economy of existence; it is a religion in which man is at once the worshiper and the god" (Durkheim 1973, pp. 45–46).

We agree with Durkheim that the human person has been sacralised but are (like H. Joas 2013) less willing to agree with the idea that a sort of religion has been established around it. As Durkheim himself was aware, religion requires institutionalisation and the systemization of a set of "beliefs and practices relative to sacred things" (Durkheim 1995, p. 44), and this cannot be said to have existed around the human person at the time when Durkheim was writing, or indeed in the present day.

This sociological "urbanisation" of the sacralisation of the person continues in the work of Joas (2013), who picks up the glove thrown down by Durkheim in 1898 and analyzes the development of the phenomenon and its persistence in contemporary societies. He begins with a dialogue with the exhaustive reflections of Foucault (1977) and Beccaria (1995) on the evolution of forms of punishment. For Joas, the key to understanding the presence of such forms in modern societies lies not so much in the need to discipline criminals (as argued by Foucault (1977)) but rather in the defence of the human being as a transcendent object: "In the history of criminal law, the worst crime has generally been that which violates the sacred core of the community. So it seems reasonable to trace changes in the penal system back to changes in the understanding of the sacred. This is why I refer to the alternative interpretation proposed here as the "sacralization of the person". From this perspective, the reforms of penal law and penal practice, and the rise of human rights in the late eighteenth century, are the expression of a profound cultural shift in which the human person became a sacred object". (Joas 2013, p. 49).

Thus, in contemporary societies, the person has become an object of veneration. As stated by Joas (2013) and Durkheim (1973), attacks on persons are among the worst crimes that can be committed. This begs the question of whether there are sacrificial representations based on this new "sacred thing". We believe that there are. Thus, to complete this look at the course of sacrifice and record its presence in modern, secular societies, we briefly analyze two examples of situations that have hit the civilian population in the past few decades: *the 9/11 attacks and the COVID-19 pandemic*. We believe that certain actions taken by groups in society such as fire-fighters, police officers and volunteers in the first case and healthcare workers in the second can be associated with a clear self-sacrificial dimension focused on the sacredness of the person. This does not mean that sacrifice can only occur nowadays in scenarios of social stress, but merely that the two examples set out here come from such scenarios. Clear differences can be seen between the two cases, but we focus here solely on the aspects that they have in common.

4.2. Self-Sacrifice and the Sacredness of the Person

In their article "The Culture of Sacrifice in Conscript and Volunteer Militaries" (2016), Richard Lachmann and Abby Stivers state that the Vietnam War was a landmark event in transforming the reasons why US soldiers earned military distinctions. Previously, honour had been associated with taking enemy lives, but now it comes to be attained mainly for saving the lives of fellow soldiers and protecting those of civilians. The ultimate expression of this new discourse comes when soldiers lay down their lives to save others, i.e., when they sacrifice their lives for a sacralised greater good. Soldiers still kill people in the course of their duty, but this shift from "offensive heroism" to "defensive heroism" (Lachmann and Stivers 2016) is a clear sign of the impact and capacity for the influence of the sacralization of the person in contemporary societies. This case study shows a clear example of the transition from ritual sacrifice to self-sacrifice, where the tragic–heroic narrative shift into one focused on the exchange relationship between something that is given like a gift and something that is relinquished in the process of getting a greater good. *We think that this shift from offensive to defensive heroism can be applied to the actions developed by 9/11 rescuers and by healthcare professionals during the COVID-19 pandemic, because the main aim of the actions developed by these professionals is to save the life of other people, even if it means to relinquish, to sacrifice, their own life.*

In the same way, *the action of a soldier, firefighter, police officers or healthcare professionals who sacrifice their lives to save others* can be seen as clearly linked to the development of specific professional ethics and the functions associated with these jobs, which include sometimes risking one's life. In this sense, we think that it is interesting to say that in a philosophical discussion about sacrifice we can identify a clear tension between the supererogatory and demandingness understanding of this social fact (Van Ackeren and Archer 2018; Urmson 1958; Benn 2016; Dorsey 2013). "Supererogatory actions are characterized as actions that are morally good, but not morally required, actions that go beyond the call of our moral obligations" (Dorsey 2013, p. 355). Demandingness actions, on the contrary, would be actions morally good and, at the same time, morally required. James O. Urmson offers an interesting example of this tension when states: "We may imagine a squad of soldiers to be practicing the throwing of live hand grenades; a grenade slips from the hand of one of them and rolls on the ground near the squad; one of them sacrifices his life by throwing himself on the grenade and protecting his comrades with his own body (. . .) It is clearly an action having moral status. But if the soldier had not thrown himself on the grenade would he have failed in his duty?" (Urmson 1958, p. 202). As we can see, the key question here is not the act of immolation but the act of relinquishment.

Without denying the existence of such elective affinities between specific professional ethics and the functions associated with these kinds of jobs, we contend that the attitude of soldiers, firefighters and police officers during 9/11 rescue or healthcare professionals during the COVID-19 pandemic in seeking to save lives (even if it means sacrificing their own) is also directly linked to the sacredness of the person. Indeed, it is the blend of these two elements that provide an understanding of the self-sacrificial attitude of certain occupational groups in Western contemporary societies.

Throughout history soldiers have sacrificed their lives, be it for their lord, their nation (these are clear examples of a tragic–heroic narrative) or, as in the case of the UN's "Blue Helmets", to protect civilian populations in conflict zones (this would be an example of the self-sacrificial narrative, focused on exchange relationship). Both narratives live together in contemporary societies, but we think the second is actually more important for understanding not only the transition from "offensive heroism" to "defensive heroism" (Lachmann and Stivers 2016) in the militaries, but the cases (analyzed here) of fire-fighters, police officers and volunteers in the aftermath of 9/11 and healthcare professionals in COVID-19 pandemic too, because the specific underlying reasons for risking their lives in each of these scenarios reveal another dimension of the phenomenon which is much more closely linked to what society holds sacred (Roszak 2020). If military honour is nowadays earned by protecting people's lives, that is because those lives have become a treasured good.

If professional ethics are intertwined with the sacredness of the person, more groups prepared to give up their lives to save others emerge. Four cases in point are *fire-fighters, police officers, the volunteers*[6] *in the aftermath of the 9/11 attacks in New York* (Gil-Gimeno 2018, 2020) *and healthcare professionals in the COVID-19 pandemic* (in most countries around the world). What these four groups have in common is that in a context of social tension they have either risked or sacrificed their lives to save others. During the rescue efforts following 9/11, 343 fire-fighters, 85 police officers and an unknown number of volunteers died. Almost 20 years later, members of these groups continue to lose their lives as a result of their actions at that time. Almost 200 more fire-fighters have died from the after-effects of breathing in toxic fumes and atmospheric contamination at Ground Zero. An Amnesty International[7] report published in July 2020 put the number of healthcare professionals who had died to date in the (still ongoing) COVID-19 pandemic at around 7000 worldwide[8].

Some actually interesting analyses point out that "religious capital" has become a central factor in coping with COVID-19 (Seryczyńska et al. 2021). The same happened during the 9/11 rescue and the commemoration ceremonies raised on this tragedy (Riley 2014). These type of stressful, traumatic and unexpected social experiences produce some answers focused on the necessity for: "peopled with god-like heroes; generative of myth, new interpersonal rituals, but also iconic circulations of familiar imagery, and it has been haunted by a relentless search for both the blame and the salvation of charismatic authority" (Alexander and Smith 2020, p. 264), that is to say, that denote a clear transcendent, religious background. Concepts as salvation, charisma, ritual, are clearly linked to sacralization processes. Our point is that people and societies continue in searching and articulating this kind of answer for solving some of the problems of their everyday life, and, for it, develop some sacrificial (in this case, self-sacrificial) forms. *In this sense, sacrifice or self-sacrifice remains as one of the main mechanisms throughout one thing, object, being, and so on, becomes sacred.*

Some social practices as, for instance, the applause ritual to healthcare professionals in Spain (and in other countries around the world) during the first months of COVID-19

Pandemic are social practices directly linked with the (post) heroic status adopted by these professionals. This is what guaranty the persistence of sacrifice in modern societies. This (post) heroic status is developed as a result of their self-sacrificial role played during the pandemic event. The same happened (and still happen) with police officers and firefighters in the city of New York. This last idea can be observed through another form of worship, in this case, less ritualized. In United States delivery trucks, vans or private cars, we can still see statements as: "To our heroes", "Serving those who Serve", and so on. A great part of these slogans are devoted to soldiers, but the number of these vehicles that are devoted to 9/11 rescuers is not minor.

The question we want to highlight is that the existence of these two worship practices (introduced here just as examples), and others that we can observe, are a consequence of a previous self-sacrificial exercise, that, at the same time, is based on the idea of the sacredness of the person. The idea of sacredness of the person—understood as a secular religiosity—and the self-sacrifices made in their his/her name produce the rise and development of a set of religious practices throughout religious life which is constantly renewed in modern societies.

The actions made by 9/11 rescuers and healthcare professionals during the COVID-19 pandemic are two clear examples of how Western societies activate to protect human life as one of its greater treasures by making self-sacrificial practices. One might assume that for healthcare workers, police officers and fire-fighters placing their lives at risk is just part of their code of professional ethics, as it is in the case of the military, but that raises the question of why this is so. Is it also part of the code of ethics of volunteers? Codes of professional ethics *per se* clearly do not suffice to provide an understanding of why these groups take the actions that they do. It is necessary to add meaning, a consecrated underlying value: the sacredness of the person on the terms set out by Durkheim (1973) and by Joas (2013). This operates in two directions: by inducing them to act in situations where the lives of civilians are under threat and, if necessary, by engaging a *pars pro toto* sacrificial mechanism in which the protection of the *personne humaine* in the abstract sense (Durkheim 1973) takes precedence over the life of any specific human person. It is worth considering this second meaning briefly: social emergencies call for a sacrifice in which part of society (the victim, or in this case the aforesaid groups of specialists) offer their lives as a gift, relinquishing them and immolating themselves for the whole (the sacrifier, or in this case civil society as manifested in the persons trapped in the World Trade Center or infected by the COVID-19 virus). The replacement mechanism developed is evident, but so is the sacred thing: the human person.

With respect to the function of sacrifice, i.e., consecration, is concerned, the sacrifier and/or the victim are moved to act by the idea of self-sacrifice or altruistic sacrifice (Schiller 1845; Dalferth 2010), though in this case it is secularised and focused on the sacredness of the *personne humaine* seen as a mundane religiosity rather than on the love felt by God the Father for his children as set out in the perspective of the "historic religions" (Bellah 1969).

As far as the development of sacrifice is concerned, some of the features developed during the Axial Age persist, such as the internalisation or spiritualisation of sacrifice and the blurring of the distinctions between sacrifier and victim, because the fact that it operates a transition from ritual to self-sacrifice means that they both begin to play the dual role of giving up lives and reaching the effects of that action through the consecration previously reserved solely for the sacrifier. In this scenario of internalisation or individualisation of sacrifice, the victim is consecrated through the action of sacrifice and can become a secular hero or a martyr of religiosity focused on the sacredness of the person. The role of the sacrifier becomes less central in regard to the first function, but he/she is still the basic recipient of the benefits of the action of sacrifice, because although the victim takes on the status of sacredness his/her act is aimed at protecting the group or a greater good. This does not mean that the sacrifier gives up the role of offering up lives to the sacred thing. The fact is that the process is articulated as one of transfer, of delegation, in which the victim acts as a representative of the sacrifier. For that delegation to be considered a sacrifice there must be a correlation between the two, in which the victim (in carrying out his/her action) is actually acting as a representative of the sacrifier group and not as a deviant individual unconnected to the sacrifier. That correlation can be seen clearly in the two examples presented above.

Even so, it must be said that in the modern period this blurring of the distinctions between sacrifier and victim also extends to the sacred thing. The human person, in general, is sacralised but the victim offered up in sacrifice is (or rather continues to be) also a human person, but a specific one. It could be argued that what is protected is human life on a general rather than an individual level, as stated by Durkheim (1973), but that does not resolve the clear tension, and even paradox, that arises when the victim and the sacred thing are one and the same. This paradox can be summarized in the following statement: "In part this is because there may be in such cases special sources of plurality and incommensurability of values, because the conflict is likely to be between something that is valued by a social group, and something that is valued particularly by an individual who has to consider self-sacrificing" (Tessman 2018, p. 376). The price paid for saving human lives (sacred thing) is, sometimes, a life itself (sacred thing). This is the reason why Weiss speaks about self-sacrifice in these terms: "the highest form of sacrifice is self-sacrifice, the deliberate acceptance of a course of action, entailing the loss of one's life (. . .) In self-sacrifice death is not chosen. Rather it is accepted, submitted to as a consequence of an effort to reach something else" (Weiss 1949, p. 80).

On the basis of the previous paragraph, this paradox reveals the transcendent and ambivalent dimensions linked to sacrifice. As pointed out by Burkert (1983) or Heyman (2007) through the experience of sacrificial killing (of self-sacrificial killing too): "One perceives the sacredness of life; it is nourished and perpetuated by death" (Burkert 1983, p. 38). Here, we can observe the important role played by sacrifice as one of the main means through the sacred gain a voice, through the sacred can be articulated. In this sense, either in the shape of relinquishment to a physical or moral good, or thanks to giving the life itself for achieving a greater good *the sacred needs for sacrifice* Sacrifice places human beings in front of the transcendence mirror.

One of the most important features of modern life is their "ability for self-correction" (Beriain 2005, p. 7). As we observed along this work, this provokes the emergence and cohabitation of several narratives in constant *dynamic tension* (Beriain 2020). In the introduction, we spoke about the cohabitation of sacrificial–ritual narratives and self-sacrificial narratives in the scholar studies on sacrifice. We think this cohabitation can be extrapolated to overall social life. Sacrifice understood in the light of ritual perspective is a violent act (here is not important if this violence is focused on the control of violence itself or, on the contrary, is a clear example of triggered violence). However, if we approach sacrifice in the light of self-sacrifice, it loses a great part of its violent dimension (at least, in which respect to the underlying logic of action), and focuses on—as we have seen—an exchange relationship based on gift and relinquishment. In this scenario, sacrifice develops a more performative and symbolic way and a less violent role (focused on immolation). Both narratives live together, *interacting and clashing*, as Max Weber reveals through their "new warrior Gods" metaphor.

In fact, the two examples we briefly analyzed in this section share some elements of these two narratives. Ritual narrative is depicted by the physical (not symbolic) life sacrificed of Police Officers, Firefighters, Volunteers and Healthcare professionals. Self-sacrificial narrative can be observed in the acts of relinquishment and gift involved in these lives delivered. Despite the two narratives living together in the examples analyzed, the cases of 9/11 and the COVID-19 pandemic reveal a stronger presence of self-sacrificial narrative than ritual, that is, in Bröckling's terms—in this case—the "chosen self-sacrifice" is most important rather than the "not chosen by oneself victim" (Bröckling 2020, pp. 230–31). In fact, the underlying narrative in these two examples is clearly the self-sacrificial one.

5. Conclusions

In this paper, we show that sacrifice remains present in today's societies. Sacrifice has not been diluted in the magma of history, but as a social fact, has adjusted its logic of action to the different scenarios in which it has developed. In this sense, in our work, we analyzed three changing dimensions with regard to ritual sacrifice: firstly, *a clear spiritualization or individualization of the practice that emanates from the transition to Axial Age religiosity and then the gradually secularized version of sacrifice. This generates a transition from ritual sacrifice to self-sacrifice.* As observed, The self becomes the subject of action, i.e., the person in charge in deciding if relinquishes him/her-self for a greater good or, in cases of force majeure, if gives willingly her/his life. Secondly, *we can observe a shift in the underlying logic of sacrificial practice from immolation to relinquishment.* In this sense, the main feature of self-sacrifice is the exchange relationship focused on what is relinquished and what is given as a gift. In the cases briefly analyzed in the last section, even when human life is given as a gift, this logic of action prevails. The good we relinquish is offered in the exchange. Through this exchange, the sacred thing is fostered. The sacred thing that we introduced in this work is the person. By contrast, the underlying logic of ritual sacrifice is immolation for establishing communication between mundane and supramundane realms. Thirdly we can observe *a cohabitation between historic and secular forms of religiosity.* On the one hand, in contemporary societies remain present the narrative of historic religions (Bellah 1969). These forms of religiosity develop around the idea of the existence of a chasm between mundane and supra-mundane orders. However, on the other hand, and at the same time, a new religious-secular narrative appears that gets behind in the sacralization of mundane questions like the nation, civil life, the person itself, the childhood or nature, among others. In the last section, and for illustrating the persistence of sacrifice in contemporary societies, we introduce a short analysis of one type of secular religiosity—sacredness of the person—that develops self-sacrificial forms. We briefly analyze two: the actions developed by 9/11 rescuers and by healthcare professionals during the COVID-19 pandemic. *In the terms we analyzed it with, sacrifice's persistence in modern societies entails three significant social dynamics: secularization of religiosity, individualization of religious and sacrificial practices, and a transition from immolation to relinquishment and gift with respect to sacrifice.*

In the first pages of this work, we made reference to the notion of Donald's *principle of conservation of gains*. In their course of action, societies have incorporated new elements that live together and interact with those typical of previous contexts. Each step taken in social life generates an increase in social complexity, requiring answers adapted to this increasing complexity. The future is not a *tabula rasa* from the past but a dynamic tension between the old and the new in the present. The same happens with the different mediations between sacred and profane things, they are constantly changing. The *principle of conservation of gains* embodies sacrifice through the dynamic tension existing among three historical dimensions of this social fact: *the ritual pre-axial*, the *axial–spiritual* and the *modern cognitive–instrumental*, the latter focused on gift exchange and relinquishment. These three dimensions do not perform in an independent way, but in an interdependent one and they appear in hybrid variations like those that have been briefly shown in the examples of 9/11 and the COVID-19 pandemic. The social evolutionary analysis of Donald leaves the door open for the development of persistent processes of emergence and re-emergence of social action logics and, therefore, of sacrifice action logics. *Self-sacrifice based on gift and relinquishment is the underlying logic in the cases analyzed, but it is not the only narrative rising around sacrifice in contemporary societies.* We think this narrative is one of the most important today, but we also think that it is not the only one.

Following Donald, our proposal tries to avoid a teleological bias. Dynamic tensions among the three dimensions of sacrifice—and also between the sacred and the profane—symbolize just the opposite. We do not defend a finalist view of sacrifice, one based on the existence of several stages. In fact, we tried to avoid conjectures like: "when a society achieves their most modern expression, it is less violent and more peaceful", which seek to eradicate sacrifice from modern social life, because they understand it as a past vestige, as a remainder of what we have been, but not of what we are. *The principle of conservation of gains* tells us just the opposite: Social facts remain and experiment with metamorphoses. In this way, for understanding them, it is necessary to take into account these two dimensions of social change and the constant dynamic tensions experimented by the old and new gains.

Author Contributions: J.G-G. and C.S.C. have jointly developed all the paper. Both authors have read and agreed to the published version of the manuscript.

Funding: "This translation was funded by Social Changes Research Group Public University of Navarra". This research was funded by the National Project "Variedades de la experiencia creativa y modelos de sociedad" (REF: CSO2017-85052-R) granted by Ministerio de Economía y Competitividad (Spain).

Institutional Review Board Statement: Not applicable.

Informed Consent Statement: Not applicable.

Data Availability Statement: Not applicable.

Conflicts of Interest: The author declares no conflict of interest.

Notes

1 In this sense, we can read an interesting paper: Belmonte (2020): "Phenomenology of resentment according to Scheler and Girard in light of sloth in Thomas Aquinas' Summa Theologiae". *Scientia et Fides* 1: 221–42. doi:10.12775/SetF.2020.002, that linked R. Girard's work to Thomas de Aquinas thought.

2 In this sense, it is important to say that we can find three theological perspectives: 1. Docetism, linked to the idea of Jesus has only divine nature. 2. Adaptionism/Arrianism, only human nature is adopted by the divine person 3. Hypostatic Union, Jesus was one person with two natures, divine and human.

3 It is important to highlight the name of the supernatural is clearly axial: *logos*, "the word".

4 This could be argued about at great length.

5 The etymology of the word can be traced back to the root meaning of "witnesses".

6 This group is added because it was one of the most prominent in the rescue works made in Manhattan following the 9/11 attacks, not for being an "occupational group".

7 https://www.amnesty.org/es/latest/news/2020/09/amnesty-analysis-7000-health-workers-have-died-from-covid19/ (accessed on 8 March 2021)

8 Supporting this idea, we introduce two statements linked to the sacrificial dimension of firefighters in 9/11 attacks and healthcare professional during COVID-19 pandemic. Firstly, the words of Rob Serra, surviving firefighter in the rescue of Twin Towers: "I do remember thinking that this is probably going to kill me". Statement removed from: https://www.nbcnews.com/storyline/9-11-anniversary/9-11-first-responders-begin-feel-attack-s-long-term-n908306 (accessed on 8 March 2021). Secondly, we introduce this little extract of an opinion piece writes by Norma Torres, member of US Congress (California): "As a nation, we have always honored the men and women who make the ultimate sacrifice, and we should do the same for our medical heroes too. That's why I introduced the Frontline Heroes Act last week. [. . .] We have never failed heroes who sacrificed their lives before, and we're not going to start now. Americans are grateful for the sacrifices that healthcare workers are making." Statement removed from: https://www.theguardian.com/commentisfree/2020/oct/09/frontline-medical-workers-coronavirus-support-congress (accessed on 8 March 2021).

References

Adorno, Theodor W., and Max Horkheimer. 2002. *Dialectic of Enlightenment*. Standford: Stanford University Press.

Agamben, Giorgio. 1998. *Homo Sacer: Sovereign, Power and Bare Life*. Standford: Stanford University Press.

Alexander, Jeffrey C., and Philip Smith. 2020. COVID-19 and Symbolic Action: Global Pandemic as Code, Narrative and Cultural Performance. *American Journal of Cultural Sociology* 8: 263–69. [CrossRef]

Aron, Raymond. 1944. L´avenir des religions séculières. *La France Libre* 8: 210–17.

Assmann, Jan. 2009. *The Price of Monotheism*. Standford: Stadnford University Press.

Bataille, George. 1992. *Theory of Religion*. New York: Zone Books.

Bauman, Zygmunt. 1991. *Modernity and the Holocaust*. Cambridge: Polity Press.

Beccaria, Cesare. 1995. *On Crime and Punishments*. Cambridge: Cambridge University Press.

Belayche, Nicole. 2001. Partager la table des dieux: L´empereur Julien et les sacrifices. *Revue de l´Histoire des Religions* 218: 457–86. [CrossRef]

Bellah, Robert N. 1967. Religion in America. *Daedalus. Journal of the American Academy of Arts and Sciences* 96: 1–21.
Bellah, Robert N. 1969. Religious Evolution. In *Sociology and Religion. A Book of Readings*. Edited by N. Birnbaum and G. Lenzer. New Jersey: Prentice-Hall, pp. 67–83.
Bellah, Robert N. 2011. *Religion in Human Evolution: From Paleolithic to the Axial Age*. Cambridge: The Belknap Press of Harvard University Press.
Belmonte, Miguel Ángel. 2020. Phenomenology of resentment according to Scheler and Girard in light of sloth in Thomas Aquinas' Summa Theologiae. *Scientia et Fides* 1: 221–42.
Benn, Claire. 2016. Overdemandingness Objections and Supererogation. In *The Limits of Moral Obligation*. Edited by Marcel Van Ackeren and Michael Kuehler. London: Routledge, pp. 68–83.
Berger, Peter L. 2014. *The Many Altars of Modernity*. Boston: De Gruyter.
Beriain, Josetxo. 2005. *Modernidades en Disputa*. Barcelona: Anthropos.
Beriain, Josetxo. 2020. The Endless Metamorphoses of Sacrifice and Its Clashing Narratives. *Religions* 11: 684. [CrossRef]
Bloch, Maurice. 1992. *Prey into Hunter. The Politics of Religious Experience*. Cambridge: Cambridge University Press.
Bloch, Maurice. 2015. *In and Out of Each Other's Bodies: Theory of Mind, Evolution, Truth and the Nature of the Social*. New York: Routledge.
Bröckling, Ulrich. 2020. *Postheorische Helden. Ein Zeitbild*. Frankfurt am Main: Suhrkamp, pp. 230–31.
Burkert, Walter. 1983. *Homo Necans: The Anthropology of Ancient Greek Sacrificial Ritual and Myth*. Berkeley: University of California Press.
Casanova, José V. 1994. *Public Religions in the Modern World*. Chicago: University of Chicago Press.
Casanova, José V. 2001. Secularization. In *The International Encyclopaedia of Social and Behavioral Sciences*. Edited by Neil J. Smelser and Paul B. Baltes. Oxford: Elsevier, pp. 13.786–13.791.
Cassirer, Ernst. 1955. *The Philosophy of Symbolic Forms. Volume Two: Mythical Thought*. New Haven: Yale University Press.
Chappell, Shopie-Grace. 2018. The Cross. *International Journal of Philosophical Studies* 26: 1–21. [CrossRef]
Comte, Auguste. 2009. *The Positive Philosophy of Auguste Comte*. 2 vols. Cambridge: Cambridge University Press.
Dalferth, Ingolf. 2010. Self-sacrifice. From the Act of Violence to the Passion of Love. *International Journal for Philosophy of Religion* 68: 77–94. [CrossRef]
Derrida, Jacques. 1998. *Faith and Knowledge: The Two Sources of Religion at the Limits of Reason Alone*. Edited by Jacques Derrida and Gianni Vattimo. Standford: Standford University Press, pp. 1–78.
Detienne, Marcel, and Jean Pierre Vernant. 1989. *The Cuisine of Sacrifice among the Greeks*. Chicago: Chicago University Press.
Donald, Merlin. 1991. *The Origins of the Modern Mind: Three Stages in the Evolution of Culture and Cognition*. Cambridge: Harvard University Press.
Donald, Merlin. 2012. An Evolutionary Approach to Culture: Implications for the Study of Axial Age. In *The Axial Age and Its Consequences*. Edited by Robert N. Bellah and Hans Joas. Cambridge: The Belknap Press of Harvard University Press, pp. 47–76.
Dorsey, Dale. 2013. The Supererogatory, and How to Accommodate It. *Utilitas* 25: 355–82. [CrossRef]
Durkheim, Émile. 1973. Individualism and the Intellectuals. In *Émile Durkheim. On Morality and Society*. Edited by Robert N. Bellah. Chicago: University of Chicago Press, pp. 43–57.
Durkheim, Émile. 1995. *The Elementary Forms of Religious Life*. New York: Free Press.
Duyndam, Joachim, Anne-Marie Korte, and Marcel Poorthuis. 2016. *Sacrifice in Modernity: Community, Ritual, Identity. From Nationalism and Nonviolence to Health Care and Harry Potter*. Leiden: Brill.
Dworkin, Richard. 2011. *Justice for Hedgehogs*. Cambridge: Harvard University Press.
Eagleton, Terry. 2018. *Radical Sacrifice*. New Haven: Yale University Press.
Eisenstadt, Shmuel N. 1986. Introduction: The Axial Age Breakthroughs- Their Characteristics and Origins. In *The Origins and Diversity of Axial Age Civilizations*. Edited by Shmuel N. Eisenstadt. New York: State University of New York, pp. 1–25.
Elias, Norbert. 1968. *The Civilizing Process. Sociogenetic and Psychogenetic Investigations*. London: Blackwell. First published 1939.
Evans-Pritchard, Edward. 1956. *Nuer Religion*. Oxford: Clarendon Press.
Feuerbach, Ludwig. 1882. *The Essence of Christianity*. London: Trübner and Co.
Foucault, Michel. 1977. *Discipline and Punish. The Birth of the Prison*. London: Allen Lane.
Gil-Gimeno, Javier. 2018. Lo religioso en escenarios de estrés social: El 11-S visto desde la perspectiva de la religión civil de Robert N. Bellah. *Revista Anthropos. Cuadernos De Cultura Crítica Y Conocimiento* 248: 115–28.
Gil-Gimeno, Javier. 2020. La pervivencia del rito en las sociedades modernas y seculares. El 11-S como ritual piacular. In *Creatividad: Entre Transgresión Y Normalización*. Edited by Celso Sánchez Capdequí. Madrid: La Catarata, pp. 175–219.
Gil-Gimeno, Javier. 2021. Las ambivalencias creativas del libro: Del monoteísmo al fundamentalismo moderno. *Revista Española de Sociología (RES)* 30: 1–18.
Giner, Salvador. 1993. Religión civil. *REIS* 61: 23–55. [CrossRef]
Girard, René. 1977. *Violence and the Sacred*. Washington: John Hopkins University.
Girard, René. 2012. *Sacrifice*. East Lansing (MICH): Michigan State University Press.
Habermas, Jürgen. 1984. *The Theory of Communicative Action. Lifeworld and System*. London: Heinemann, vol. 2.
Hamayon, Roberte. 1990. *La Chasse À l´Âme. Esquisse d´ Une Theorie Chamanienne Siberienne*. Nanterre: Societé d´Ethnologie.
Hénaff, Marcel. 2010. *The Price of Truth*. Standford: Standford University Press.
Herrenschmidt, Olivier. 1989. *Les Meilleurs Dieux Sont Hindous*. Lausanne: L´Âge D´Homme.

Heyman, George. 2007. *The Power of Sacrifice: Roman and Christian Discourses in Conflict*. Washington, DC: The Catholic University of America Press.
Jaspers, Karl. 2011. *The Origin and Goal of History*. London: Routledge.
Jay, N. 1992. *Throughout your Generations Forever: Sacrifice, Religion and Paternity*. Chicago: University of Chicago Press.
Joas, Hans. 2013. *The Sacredness of the Person. A New Genealogy of Human Rights*. Washington, DC: Georgetown University Press.
Keenan, Dennis K. 2005. *The Question of Sacrifice*. Bloomington: Indiana University Press.
Lachmann, Richard, and Abbey Stivers. 2016. The Culture of Sacrifice in Conscript and Volunteer Militaries: The US Medal of Honor from the Civil War to Iraq, 1861–2014. *American Journal of Cultural Sociology* 4: 323–58. [CrossRef]
Lévi, Sylvain. 1966. *La Doctrine Du Sacrifice Dans Les Brâhmanas*. Paris: Presses Universitaires de France.
Levi-Strauss, Claude. 1966. *The Savage Mind*. London: Weidenfeld and Nicolson.
Marvin, Carolyn, and David W. Ingle. 1999. *Blood Sacrifice and the Nation*. Cambridge: Cambridge University Press.
Mathiez, Albert. 2012. *El origen De Los Cultos Revolucionarios*. Madrid: CIS.
Mauss, Marcel, and Henri Hubert. 1964. *Sacrifice: Its Nature and Functions*. Chicago: Chicago University Press.
Mauss, Marcel. 2002. *The Gift: The Form and Reason for Exchange in Archaic Societies*. London: Routledge. First published 1924.
McClymond, Kathryn. 2008. *Beyond Sacred Violence: A Comparative Study of Sacrifice*. Baltimore: Johns Hopkins University Press.
Milbank, John. 1996. Stories of Sacrifice. *Modern Theology* 12: 75–102. [CrossRef]
Nancy, Jean Luc. 2003. *A Finite Thinking*. Standford: Stanford University Press.
Nietzsche, Friedrich. 1967. *On the Genealogy of Morals and Ecce Homo*. New York: Random House.
Parsons, Talcott. 2005. *The Social System*. London: Routledge.
Petropoulou, Maria Zoe. 2008. *Animal Sacrifice in Ancient Greek Religion, Judaism and Christianity, 100 BC-AD 200*. Oxford: Oxford University Press.
Riley, Alexander. 2014. Flags, Totem Bodies, and the Meanings of 9/11: A Durkheimian Tour of a September 11th Ceremony at the Flight 93 Chapel. *Canadian Journal of Sociology* 39: 719–40. [CrossRef]
Rives, James. 1995. Human Sacrifices among Pagans and Christians. *Journal of Roman Studies* 85: 65–85. [CrossRef]
Robertson Smith, William. 1886. Sacrifice. In *Encyclopaedia Britannica*. Edinburgh: Adam and Charles Black, p. 132.
Robertson Smith, William. 1972. *The Religion of the Semites. The Fundamental Institutions*. New York: Schocken.
Roszak, Piotr. 2020. Mute Sacrum. Faith and Its Relation to Heritage on Camino de Santiago. *Religions* 11: 70. [CrossRef]
Sánchez Capdequí, Celso. 2004. *Las Máscaras Del Dinero. El simbolismo Social de la Riqueza*. Barcelona: Anthropos.
Schiller, Friedrich. 1845. *Aesthetic Letters, Essays and Philosophical Letters*. Boston: Charles C. Little and James Brown.
Seaford, Richard. 2004. *Money and the Early Greek Mind: Homer, Philosophy, Tragedy*. Cambridge: Cambridge University Press.
Szryczyńska, Berenika, Lluis Oviedo, Piotr Roszak, Suvi-Maria Katariina Saarelainen, Hilla Inkilä, Josefa Torralba Albaladejo, and Francis-Vincent Anthony. 2021. Religious Capital as a Central Factor Coping with the COVID-19. Clues from an International Survey. *European Journal of Science and Theology* 17: 43–56.
Shilling, Chris, and Philip A. Mellor. 2013. Making Things Sacred: Re-theorizing the Nature and Function of Sacrifice in Modernity. *Journal of Classical Sociology* 13: 319–37. [CrossRef]
Simmel, Georg. 2004. *Philosophy of Money*. London: Routledge.
Spencer, Herbert. 2004. *The Principles of Sociology*. 4 vols. Forest Grove (Oregon): Pacific University Press.
Strenski, Ivan. 2002. *Contesting Sacrifice: Religion, Nationalism and Social Thought in France*. Chicago: Chicago University Press.
Stroumsa, Guy. 2009. *The End of Sacrifice: Religious Transformations in Late Antiquity*. Chicago: University of Chicago Press.
Stroumsa, Guy. 2017. *The Making of the Abrahamic Religions in Late Antiquity*. Oxford: Oxford University Press.
Tava, Francesco. 2018. Sacrifice as a Political Problem: Jan Patocka and Sacred Sociology. *Metodo* 6: 71–96. [CrossRef]
Taylor, Charles. 2007. *A Secular Age*. Harvard: Harvard University Press.
Tessman, Lisa. 2018. Sacrificing Value. *International Journal of Philosophical Studies* 26: 376–98. [CrossRef]
Theissen, Gerd. 1999. *A Theory of Primitive Christian Religion*. London: SCM Press.
Tylor, Edward B. 2010. *Primitive Culture: Researches into the Development of Mythology, Philosophy, Religion, Language, Art and Custom*. Cambridge: Cambridge University Press.
Urmson, James O. 1958. Saints and Heroes. In *Essays in Moral Philosophy*. Edited by Abraham I. Melden. Washington, DC: University of Washington Press, pp. 198–216.
Van Ackeren, Marcel, and Alfred Archer. 2018. Sacrifice and Moral Philosophy. *International Journal of Philosophical Studies* 26: 301–7. [CrossRef]
Van Baal, Jan. 1976. Offering, Sacrifice and Gift. *Numen* 23: 161–78. [CrossRef]
Weber, Max. 2004. *El político Y El Científico*. Madrid: Alianza.
Weiss, Paul. 1949. Sacrifice and Self-sacrifice: Their Warrant and Limits. *The Review of Metaphysics* 2: 75–98.

 MDPI

Article

Metamorphosis of the Sacrificial Victimization Imaginary Profile within the Framework of Late Modern Societies

Angel Enrique Carretero Pasin

Department Philosophy and Anthropology, Faculty Philosophy, University Santiago de Compostela, Square Mazarelos, s/n, 15782 Santiago de Compostela, Spain; angelenrique.carretero@usc.es

Abstract: This article aims to unravel the why and the how of the imaginary profile of the emerging sacrificial victim in late modern societies. To do this, first, under the influence of the formulations proposed by the *French School of Sociology*, the nature and the functionality of an anthropological structure linked to a rituality of sacrificial victimization surviving in the historical course of western societies are investigated. Based on this, it analyzes the characterization of the imaginary paradigm of sacrificial victimization crystallized in modernity in contrast to the dominant one in the *Old Regime*. Finally, the sociological keys that would account for the unique morphology of the imaginary of sacrificial victimization that emerged in late modern societies are explored in the context of the generalization of a climate of violence that transforms any individual into a potential victim of sacrifice.

Keywords: imaginary; sacrifice; violence; rituality; collective communion; late modernity

Citation: Carretero Pasin, Angel Enrique. 2021. Metamorphosis of the Sacrificial Victimization Imaginary Profile within the Framework of Late Modern Societies. *Religions* 12: 55. https://doi.org/10.3390/rel12010055

Received: 17 December 2020
Accepted: 11 January 2021
Published: 14 January 2021

Publisher's Note: MDPI stays neutral with regard to jurisdictional claims in published maps and institutional affiliations.

1. Introduction

To start with, it is worth highlighting that social sciences have not fully assumed the radicality of the *dictum* launched by Girard: "Sacrifice is the primary institution of human culture" (Girard 2012, p. 75). This work follows along the lines of this premise. The publication of *Totem and Taboo* by Freud—under the influence of Frazer—promoted a horizon to address the foundation both of the sacrificial rite and of the sacrificial component that remain latent in culture. It is well known that Freud developed the hypothesis according to which totemism would have emanated from a universal oedipal complex. In a dialog with the scholar in Semitism Robertson-Smith (1894), Freud (1988, pp. 173–209) argued that the shared meal ritual serves to fraternize. Some decades later, Lévi-Strauss (1964, p. 145) suggested something similar: totemism "founds an ethics" full of "prescriptions and prohibitions", which "seems to result from the very frequent association of totemic representations, on the one hand, with food prohibitions and, on the other hand, with exogamy rules". It seemed exceedingly surprising to Freud that totemic sacrifice should require full commitment on the part of the group. For him, the meal would represent the primeval act catalyzing religion—the crystallization of a social order supported on a communion of feelings of affection. Otherwise, evidence exists that sacrifice would have been justified in the symbolic order of primitive societies by the fact that everybody took part in that practice (Baudrillard 1980, pp. 156–62). Not by chance, its addressees were members of the royal dynasty who exercised the representation of the community and of the sacred at the same time (Frazer 1981, pp. 338–42).

Nonetheless, Mauss and Hubert's thesis is the one marking a turning point in the vision about the sacrificial cult. The fertility of their approach lies in stressing what they call "unity" of the sacrificial rite—an anthropological structure which persists beyond morphology where sacrifice materializes. Mauss and Hubert endowed the theoretical status of sacrifice with solidity, moving away from the British anthropological tradition, which they accused of being arbitrary in its premises around totemism and inconsistent in its methodological principles. They understand sacrifice as an act which fosters a break and a transformation in the participant's interiority, where *the sacrifice* comes into contact with *grace*, collecting—in return—some benefits at an individual, clan, tribe, or nation level (Mauss and Hubert 1970, p. 155). The sacrificial rite forces a communication between the sacred and the profane that uses the victim as an intermediary (Mauss and Hubert 1970, p. 244). Both spheres are scrupulously separated in everyday existence. An expiation and communion effect would simultaneously take place in sacrifice. A regenerative community redemption becomes distilled via immolation. Purity and cohesion would thus go hand in hand[1]. The religious ceremony recharges those participating in the rite with emotional energy, making them enter a "collective communion". The victim has the gift of irradiating that energy. Sacrifice "periodically renews the good, strong, terrible, and serious character which arises as one of the essential features of any social personality in the collectivity—represented by their gods. Therefore, the social norm is maintained without any danger for them, and without any decrease for the group" (Mauss and Hubert 1970, p. 248). By means of a thorough historical-anthropological journey, Mauss and Hubert leveled out an interpretive horizon so that, more than a decade later, Durkheim could underpin his hypothesis concerning the sacred; rite works with a view to reaffirm group conscience. The practicity of sacrifice will reside in its capacity to catalyze a collective space–time coexistence through the regeneration of a "communion of consciences". Durkheim (1982, pp. 303–25) used the expression "positive cult" to describe a particularity of the sacrificial rite in accordance with a transitory abandonment of the profane universe by the faithful and an opening towards the universe of the sacred. The French scholar divided the ritual typology into two aspects: "communion" and "oblation". In Durkheim's opinion, being the hypostasized and transfigured translation of society, the gods could not neglect human beings, since the survival of the former would, from time to time, depend on the latter, an inflamed "communion of consciences" which, episodically revived through ceremonial gestures, enables the community to show itself as "more alive" and "more real".[2]

More or less heterodox continuators of the trail initiated by Mauss and Hubert emphasizes that the notions of sacrifice and violence are *per se* closely linked (Bataille 1987, pp. 81–97). In terms of homeostatic balance, war would be a healthy event, throwing the swarming violence accumulated inside a group at an external enemy that becomes the object of sacrificial destruction for the purpose of demarcating the acquiescence or the distance regarding this group entity. Clastres (2004) verified that the movement of violence towards an external figure served to reaffirm an unyielding difference of the group, especially of the one that tried to counter a drift towards its splitting. Primitive societies were made for war because it is in that scenario that their essence as a society takes shape. If the enemy did not exist, it would have to be invented. In human groups that, unlike others—e.g., *The Nuer*—lack resources to arbitrate between conflicting parties (Evans-Pritchard 1977, pp. 163–203), this violent *pathos* could be projected towards the figure of an internal enemy.

In this respect, the most refined formulation is undoubtedly the one carried out by Girard (1983, pp. 9–45). Community expiation through the use of human victims had already attracted Frazer's attention in the late 19th century. He had identified it in Ancient

---

1 Purity "founds, maintains or perfects a norm, an order, a health. It is understood that the sovereign will embody them" (Callois 1996, p. 55). A sacrifice is a rite exercised upon an animal that owns something divine. The immolated victim shows an ambivalent aura; it is chosen to be sacrificed precisely for being sacred. Purity and impurity are not mutually exclusive. In primitive languages, the verb "to purify" meant both "to cure" and "to exorcize." In Rome, *sacer* designates who can or cannot be touched without being stained or staining. These two dissociated poles keep their ambivalence (Callois 1996, pp. 29–39). It is assumed that the sacrifice which consecrates the victim includes its destruction.

2 In the threshold of pre-modern societies, rite has played in favor of collective communion, helping union to defeat disunity (Balandier 1996, pp. 28–35), taking advantage of the fact that, by means of rite, "norms and values are loaded with emotions, whereas basic and rude emotions are given dignity through their contact with social values" (Turner 1980, p. 33). The hermeneutics of sacrificial victimization is framed within such parameters. Thus, purity rites, in conflict with a metaphorical contamination, focus on fixing barriers to chaos, establishing "a link between order and disorder, being and non-being, form and shapelessness, life and death" (Douglas 1973, p. 19), insofar as they eagerly pursue to exorcize the indefinable as an anathema of evil nature, hence the fate suffered by figures labeled as witches or misfits people, illustrations of evil (Douglas 1978, pp. 134–50). For that reason, in order to avoid a profanation gesture in the sacrificial ceremony, the victims should rid themselves of any stain the same as the priest, who is put on a level with the height of purity (Douglas 1973, pp. 73–74).

Greece, in the Saturnalia of the Roman Empire and in the primeval Aztec culture, revealing the contradictory divine halo which is inseparable from the men and the women chosen for sacrifice (Frazer 1981, pp. 651–66). Nevertheless, Girard tried to unravel the motivation behind sacrificial victimization by virtue of how a group fixes a violence born from a "mimetic rivalry" on somebody—a victim—who, if not within reach, would threaten their integrity. According to the force-idea, one can only deceive violence by offering it a substitutive bait in return. Thanks to sacrifice, internal violence is appeased, and the outbreak of brawls is canceled, thus activating a beneficial violence which offsets another harmful one. Wherever a set of tensions prevailed, it resulted in the emergence of a community unified around the unanimous hate aroused by one of its members.[3] This confirms the existence of the "scapegoat"—foundational martyr-myth which gives strength to the group, the addressee on which the interwoven feelings of hate converge so as to preserve the social body, an enemy of the community that favors intragroup friendship at the cost of his curse and punishment.[4] The role played by the sacrificial rite is summarized in taking man away from violence to protect him there from.[5] Girard's approach posed a remarkable anthropological challenge: "that of the sacral and victimizing nature of every culture, which uses a plethora of ways to mask that sacrality and *numinosity* and, of course, the historical victims and slaughterhouses, but which could neither exist nor survive without all of that" (Jiménez Lozano 1999, p. 13).[5]

Nonetheless, the theoretical line inspired in the *École Française de Sociologie*, overlapped on the theses of Mauss and Hubert and whose dissemination organ was *L'Année Sociologique*, did not succeed in becoming consolidated as the hegemonic paradigm. This made the theme of sacrifice head towards a certain degree of untimeliness. With the intention of revisiting the deepest structures of collective feeling, the spirit of this *École* was rescued years later by the *Sociology of the Sacred*, driven from the *Collège de Sociologie* (Bataille, Callois, Leiris ...) and the heterodox magazine *Acéphale*. In recent times, an effort has once again been made to recover the significance of the Maussian (Durkheimian) perspective around the sacred ritual[7] and, in parallel, to activate the socio-anthropological implications derived from the modulation of the sacrificial phenomenon in late modernity.[8] In this context, the analytical hypothesis suggested here is one of a transition in the profile of the imaginary

3 This is in concomitance with what has happened in a family constellation based on a deteriorated communicative relationship that favors the outbreak of schizophrenia in a son/daughter on whom is fixed the status of "scapegoat" that unifies the family system, as was highlighted first by antipsychiatry and then by the Palo Alto School.

4 "Witch hunt" has been interpreted as a contemporary illustration for the invention of a common enemy that reinforces the sentiment of group unity along the lines of Durkheim's approach (Bergesen 1978).

5 Something similar was defended by Canetti (1983, pp. 51–56) in his portrait of the "harassing masses", whose anger is unleashed in public executions, where the crime upon the victim necessarily has to be collective, even if the hangman was its executive arm.

6 Not everybody shares Girard's conception of sacrificial anamnesis. Caillé (2000) has criticized him for having hypostasized mimetic desire as a cause of the sacrificial rite, obviating a previous symbolic reference crisis whose deficit would catalyze sacrificial victimization. Following Mauss, he situates the gift and the counter-gift as the essence of any social relationship, including the sacrificial one. Sacrifice is interpreted as an agonistic gift. This discrepancy is not new. He separates the thesis of Voltaire and De Maistre. "A large amount of evidence demonstrates that the first human victims were culprits condemned by the laws; since all nations have believed what, according to Caesar's reports, was believed by druids; that the punishment of culprits was very pleasant for the divinity. The ancients thought that every crime committed in the State related to the nation, and that the culprit was consecrated and offered to the gods until, due to the shedding of blood, both he himself and the nation could be freed" (De Maistre 2019, p. 134). The Savoyard understood the sacrificial bloodshed as the reparation of an original evil concentrated on an innocent except for the fact that, unlike Girard, De Maistre conceives this act as a historical driver. Jesus of Nazareth matches this profile, the same as Louis XVI later. In turn, Hénaff (2002, pp. 251–67) envisages sacrifice as an offering to an invisible addressee in whom we wish to cause an anti-utilitarian response—the donation of *grace*. In his view, Girard's pessimism would be betrayed by his faithfulness to Jesus of Nazareth's sacrificial hermeneutics, making the mistake of confusing the fact of being the object of sacrifice with a victim and the sacrificial act with a victimization *tout court*. Girard (2012, pp. 75 and ff.) identified a concomitance between the characterization of sacrifice in Vedic religions (*veda* designates the science of sacrifice as the principle of this religion)—explicit in the book *Rig-Veda* and unraveled in its second stratum, *The Brahmana*—and Jesus of Nazareth's crucifixion. However, for Hénaff, the substance of sacrifice points at a metaphysical context, a cosmic equilibrium between nature and culture which periodically reestablishes a meaning order for a human group. Without discrediting the proposals made by Caillé and by the same author together with Godbout (Godbout and Caillé 2000) as well as by Hénaff, considering the objective sought in this work, it seems advisable for us to follow the line advocated by Girard.

7 The contributions made by Delgado (2001, 2008), Páez (2002), Bergua (2007), Cerruti (2010), Juan (2013), and Lorio (2013) are of special interest.

8 Enriching contributions in this regard include those of Caillé (2000), Godbout and Caillé (2000), Hénaff (2002), Beriain (2007, 2009, 2017), Magdalena (2015), Le Breton (2017, 2018), and Delgado and Martín Delgado and López (2019).

associated with the sacrificial victim during late modernity with respect to the paradigmatic victimization of modernity whose genesis and outlining are specified below.

2. Method

A socio-historical genealogy serves to explain the structural transformations that trigger the emergence of a latemodern sacrificial paradigm, laying the emphasis on the factors driving it and, at the same time, on a socio-hermeneutics of the significance frameworks where a consequent intersubjectivity is forged. To that end, this work took as a reference the examination of the everyday implications induced at this level by the socio-anthropological analyses proposed through the selection of various authors whose focus of attention has been to highlight the disintegration suffered by the symbolic fabric meant to guarantee an intersubjective universe lived in common. This phenomenon—in turn caused by the establishment of modernity—became hyperbolically exacerbated within the context of latemodern social formations. Bearing in mind the above, a scrutiny is likewise performed of the reasons favoring a turn that marked a tipping point in such social formations when managing the violence harbored inside them and how that key turn is reflected in a modification of the profile corresponding to the most usual victimization praxis.

3. Results and Discussion

3.1. *Synopsis of the Sacrificial Victim Imaginary in Modernity*

3.1.1. Basic Principles

Let us start from the idea that the western secularizing process did not extinguish the differentiation between profane and sacred domains. A more conceptually refined analysis about the nature of modernity betrays a maintenance of this differentiation, even though the substantivity of the sacred sphere is redefined under a forcibly immanent expression.[9] Thus, two fundamental *numens* appear in the Modern Age:

a. Nation-State: after the dismantling of the feudal power system, the nation-state arises as the organ meant to monitor intraterritorial disagreements (Elias 1989, pp. 229–53), becoming institutionalized as the representative of "legitimate violence" within a geographical perimeter (Weber 1993, pp. 43–44) and endorsed as a central body in charge of collective integration.

b. Productivism: a "metaphysics of production" (Baudrillard 1996, pp. 53–61) in alliance with the industrialism unleashed by the bourgeoisie backed by the enthronization of the progress category, a secularized equivalent to divine providence (Bury 2009, pp. 32–40), and supported on the conviction that constant production growth will have positive effects in terms of public usefulness, well-being, and widespread happiness (Polanyi 1997, pp. 247–65).

As a reinforcement of their inviolability, both *numens* boast about the use of ceremonies clad with an aura of solemnity and subject to a marked time periodicity. Their role consists in reaffirming a unanimous commitment to sacrality that they carry with them in the impulse to an acquiescence around them that re-establishes the vitality of the collective tie. In return, those *numens* demand a sacrifice of individual consciences to their *desideratum*. The functional significance of both *numens* simultaneously clarifies the

[9] The immanence/transcendence code, the foundation of religion, goes beyond the opposition-based semantics of mundane and *supramundane* order (Luhmann 2007, pp. 49–100). Once this code has been recognized, the umbilical cord that links emotional community and sacrifice becomes evident in the foundational speeches of the North American nation in tribute to its martyr heroes out of consideration for the redemption of a national unity (Bellah 1967) as well as the aggrandizement of patriotic sacrifice in the interest of the survival of the nation, contemplated as a mystical *supraindividual* brotherhood that guards the memory of their sacrificed ones in war (Anderson 1993). For this reason, we agree with Gutiérrez Martínez (2010) on the urgent need to undertake an epistemic rethinking of religiousness assigning importance to the meaning of belief, in an elastic sense, within the heart of modernity.

socio-anthropological codes involved not only in a "metamorphosis of the sacred"[10] but also, more importantly, in their classification within a constellation of sacrificial plexuses.[11]

The utilization of the word "metamorphosis" refers here to an expressiveness of historic-cultural forms where something essential and underlying remains. The elucidation of that something would perhaps take us to the notion of archetype, conceived as a "dynamic structure" which, subject to modification, shapes a cultural model (Durand 1982, p. 57), or to that of "semantic basin", which gives new meaning to an immemorial, original, and primordial background adhered to nascent national currents (Durand 1996, pp. 85–136). Others have opted for a related word—"transfiguration"—to point at something similar (Maffesoli 2005), turning *formism* into a hermeneutics that sheds light on the changing appearance of things *por ricorso* to the invariant (Maffesoli 1993, pp. 79–96). A re-signification of the sameness of "forms" after the diversity of "contents" is indeed sought (Simmel 1986, pp. 11–37). Whatever the term chosen, it evidences the survival of a sacrificial structure, together with a consequent victimization phenomenology, in social formations where *prima facie* the umbilical cord with religion would have been short-circuited.

3.1.2. Sacrifice and Dissent

Modulation of sacrificial victimization: the assimilation of dissidence with evil is as little original as the formulas devised to control it (Carretero 2016).[12] Control practices are intensified with the arrival of the Modern Age, though. The dissident difference, a readapted format of evil, is labeled as enmity against a society objectivized in the State. Evil comes to be identified with an imperfection punished by the ontological totalitarianism expressed in the modern *numens* (Maffesoli 2002, pp. 73–109). Dissidence is assessed according to a correct or anomalous fit in the roles tolerated from the prerogatives urged by a modern evolutionary logic handled in time to a functional differentiation of its political and economic subsystems (Luhmann 1998, pp. 71–98). The imaginary profile of the sacrificial victim in modernity stems from a fault in systemic adequacy with respect to the directive of its *numens*.[13] With the Declaration of Rights of Man and of the Citizen, except for the countries whose criminal code accepted the validity of death penalty, the self-affirmation of the "communion of consciences" around "the social divine" (Durkheim) appeals to a standardized exclusion strategy under a space-time lockup as a sacrificial canon. Goffman (1972) and Foucault (1994) showed the true relevance of a de-culturing type of rituality which annuls the self, parallel to the confinement of individuals in the modern—"total"—disciplinary institutions, designed with a focus on isolation, i.e., psychiatric hospitals and prison facilities.[14]

Perhaps nobody dissected *la coupure* in the strategies of political power from the *Ancien Régime* to the consolidation of modernity better than Foucault (1994, pp. 11–37). His description of the punishment imposed upon Damiens—a figure who supposedly carried

---

10 See Estruch (1994), Prades (1998), Sánchez Capdequí (1998), Delgado (1999), Carretero (2003), and Sánchez Capdequí and Carretero (2006) in this regard.

11 Suggestive readings include Hénaff (2002), Beriain (2007, 2017), Casquete (2007), and Díez de Velasco (2008).

12 So a sacrificial exorcism focused on difference eases the tensions inside a group and strengthens its stability. For example, in the Greek tragedy, Antigone, Phaedra, or Orestes are sanctified and, at the same time, sacrificed for having committed sacrilege against the established *numens* against a Law-founding "mythical violence" (Benjamin 1998, pp. 44–45). Euripides' play shows characters misgoverned by a demonic *thymós* that leads to their doom—the affront to the city spirit (Dodds 2006, pp. 171–94). Rome gives way to the show of the Roman circus as a demonstration of fear about deviationists, misfits, and heretics. In the Middle Ages, the detachment of behavior from the expectations of the normative order is the object of inquisitorial repression. The accusation of witchcraft is its clearest illustration (Caro Baroja 1970, pp. 183–282), (Duvignaud 1979, pp. 175–87). Philip II and Luther, Catholics and Protestants, did agree on this point. During the Contemporary Age, the figure of the libertarian concentrates the reprisal of society against an anomic difference (Duvignaud 1990, pp. 125–45). In general, a mis-shaped social image is penalized as an a-social or an anti-social alarm, transforming the individuals who transmit it into "scapegoats" as a personification of evil, attacking group harmony, and questioning their certainties (Ferro 1984, pp. 373–81). The treatment given to the Jewish community was due to similar causes (Rusche and Kirchheimer 1984, pp. 15–24). Sacrificial exemplifications which embody the ancestral fight waged between good and evil, order and disorder.

13 Its dystopic face isAuschwitz with Eichmann as the perversity in its effectiveness (Arendt 2012).

14 It is revealing that the only shelter for the implementation of the *communitas*—an authenticity connection that was unfathomable for society's systemic-structural organization—has been the field of mysticism (Turner 1988, pp. 143–44), (Foucault 2006, pp. 256–57), or that those "a-structural" symptoms which interfered with the technical efficacy of modern society were outlawed as unacceptable (Duvignaud 1979, pp. 176–77).

out an attempt against the life of the French king in 1757—represents the vengeance of the State's legal bodies towards the corporeality of those who infringed upon a sacralized norm at the dawn of the modern world.[15] The torture inflicted upon Damiens reflects the sacrificial execution that was paradigmatic of the *Ancien Régime*. His status as a sacrificial victim has to do with his aggression against the embodiment of the sacred in a human figure according to the dominant metaphysical-political worldview. A sacrality into which the proclamation of inviolability in the unity of the collective body would be translated. An inviolability protected by a "pastoral power"—gestated in the ideological orbit of Christianity—is subsumed by the *governmentality* of the Modern State (Foucault 2006, pp. 293–326). The graduation of the punishment is toned down between 1760 and 1840. In parallel, a qualitative increase of disciplinary power takes place in the heart of a set of modern institutions which, driven by a disguised normalizing purpose, resort to the legitimacy granted to some emerging expert fields of knowledge that enjoy the approval of a new "truth regime" (Foucault 1992, pp. 175–89).[16]

Dictated by the postulates of industrialism, the reprisal unleashed upon another hypothetical Damiens is oriented towards the field of subjectivity. The treatment given to madness, with an uncomfortable accommodation in the carving of the *dictum* of modern *numens*, follows a course resembling the violation of the legal framework (Foucault 1967, pp. 76–175). Both realities—madness and crime—illustrate the threat of something that is *a-social*, a shadow of "undifferentiated heterogeneity" opposed to any rule (Bataille 1993, pp. 27–32) and untranslatable from modern systemic patterns. For that reason, the dysfunctional a-social ones, those who do not make pacts or owe a commitment or debt to the aforesaid *numens*, are expelled from the community via their isolation; in sum, they are sacrificed by means of reclusion.[17] The State is the organ exclusively entrusted to play the role of *sacrificer*, responsible for purging a hint of anomaly that can be put on a level with a disorder which has ever since been seen as excrescence. The modern victimization imaginary profile is edified on these pillars.[18]

3.2. *Metamorphosis of Sacrificial Victimization in the Imaginary of Latemodern Societies*

3.2.1. Main Sociological Factors Contributing to the Mutation of Sacrificial Mythology

The analytical hypothesis proposed by this metamorphosis makes it necessary to address the following aspects:

a. Self-affirmation for the cult of the self: favored by the rise of liberalism, the deployment of modern society enthroned the will of the self, dismantling a shared

[15] In the Middle Ages, body whipping, mutilation, and burning were usual sentences imposed upon whoever violated the sanctity of the law, especially the right of ownership, a type of sentence which continued until well into the 16th century. In England, with a population of three million inhabitants, 72,000 thieves were hanged during the reign of Henry VIII. In the reign of Queen Elizabeth, vagabonds were simultaneously lynched in rows of 300 to 400 (Rusche and Kirchheimer 1984, p. 20).

[16] Until well into the 18th century, the poor, beggars, and vagabonds, a population formerly recruited for wars with neither a citizenship status nor an official useful activity and consequently prone to crime, had as their destination work in the galleys, deportation in the mines of the American continent and correctional institutions, faithfully following a combined confinement and productivity logic in a protohistory of prisons (Rusche and Kirchheimer 1984, pp. 46–71).

[17] No sacrificial anathema is applied to poverty, despite its dysfunctionality. Unlike a criminal or a madman, the poor man is recognized in his individualized condition, not with the aim of subjecting him to an institutional normalization as an entirely a-social being. This is so because, in spite of having broken his bond with the productive *numen*, he has not completely broken his tie with the State. In fact, "he acquires the condition as a poor man only when he is assisted" (Simmel 2011, p. 87). His extraterritoriality is not the one of a medieval poor man—who survived attached to the theological meaning of alms; it does not imply a loss of ties with the collective unit, i.e., with the State's duty to find formulas for his governmental integration through aid or charity institutions. "However, this isolation does not mean a separation, an exclusion; instead, it entails a particular connection with the whole which would be different without this element" (Simmel 2011, p. 81).

[18] Enquiring into the origin of the condemnation of the modern era for its a-social approach, some continuators of Foucaultian biopolitics have gone further. The path drawn by Agamben (1998, pp. 151 and ff.) rethinks a sacrificial myth which, rooted in the essence of western culture, is transposed to the idiosyncrasy of modernity. Esposito (2003, pp. 30–34) has recognized the existence in the genesis of the *communitas* of a *munus*, a debt, a common duty, or an oath that carries with it a sacrifice of subjectivity. This means that the latter is no longer its own master, being stripped within an absence of binding content riddled with nihilism. The *communitas* would be an aggregate of individuals grouped together around that debt or duty where they banish themselves. Sacrificial victimization arises from an original sacrifice of the community which, because of its existence, becomes objectivized in an exterior manifestation and thus relinquishes its authenticity.

symbolic universe which underpins the community link. The disappearance of binding contents resulted from the slogan hoisted by modernity, according to which the responsibility of each person's destiny exclusively depends on himself (Bell 1987), (Bellah et al. 1989). In the absence of a functional replacement for the normative entities protected by tradition, the alternative to the exacerbation of individualized freedom generates a latent opposition of everybody against everybody else monitored by the State, in parallel to the generalization of a quartering of emotionality inside the self (Elias 1989).

b. Dismantling of confidence-building entities: complexity has enlarged its breadth. The possibilities about what the world is and might be have multiplied. This favors the arrival of uninvited guests such as excessive instability and uncertainty as an added price to contingency. Luhmann (1996, pp. 5–80) has understood that the *raison d'être* for social subsystems is the fight against complexity using systemic formulas meant to reduce it and helping make the world become simpler and commonly familiar. Confidence reduces complexity. With modernity, the advance of complexity exceeded the familiarity framework that served as the pillar of confidence in traditional societies. This confidence, based on an expectation of continuity in the behavior with others in the context of everyday interaction (Goffman 1981), came to be a risky undertaking everywhere. The formulas for confidence which depend on familiarity turned useless, their task being transferred to the systemic sphere, which does not prevent the latter from revealing more than sporadically its inability to control complexity and to encourage intersubjective communication.

c. Evolution in the control modality: the control supported on the inclusion/exclusion binomial stopped being a functional priority. Its new strategies are redirected towards a full inclusion outside which nobody can situate themselves. The enemy that needs to be fought continues to be systemic dysfunctionality, though the formula to confront it aims at a biopolitical configuration of subjectivities which is not going to require reclusion insofar as the devices that play a prominent role in maximizing control find not only discipline and punishment but also quartering unnecessary (Deleuze 1995, pp. 403–40).

3.2.2. De-Ceremonialization and Democratization of Sacrificial Victimization

It is well known that the Modern Age turned the State into the institution *par excellence* when it came to guaranteeing the representation of the social sphere, an institution that manages that sort of unacceptable praxis from the key *numens* around which the modern collective communion is structured. As seen above, the monopoly of social control rested upon the State by means of a peculiarly expiatory institutional exercise.

Thus, late modern societies redefine sacrificial phenomenology. They show a decommitment of the control by the State along with a delegation of its exercise to the will of each individual. The disarticulation of the role played by the State as a fixed center of gravity in the supervision of the social context correlates with the weakening of its institutional authority in the transmission of normativity as well as in the framing of a systemic integration freed from a moral concordance (Luhmann 1998, pp. 197–212). Consequently, institutionalized rites—vested with a halo of solemnity and oriented to reaffirming the axiomatic values of collective conscience—enter a spiral of wear and disaffection.[19] Sacrificial rites—as a subclass of institutionalized rites—break down. Until well into modernity, when such a ritual desire still remains alive, the wealth of violence accumulated inside a group, the "collective shadow" self-denied therein (Jung 1966), had been exorcized through a twofold channeling, internally by the catharsis periodically triggered in festivities and at an external level by a projection—evoking patriotism—towards warfare. The degradation

[19] "I will define ritualism—Douglas said—as an impassioned appreciation of symbolic action which materializes, firstly in the belief in the effectiveness of the instituted signs, and secondly in the sensibility regarding the condensed symbols" (Douglas 1978, p. 22).

in the authority of the common rites goes hand in hand with a de-regulation of sacrificial ritualism incorporated into the day-to-day praxis.

An enlightening approach to this de-regulation is provided by Michaud (1980, pp. 73–96), in whose opinion the social sphere has experienced a situation of ambiguity since modernity —the artifice of its maintenance without the support of an adherence, building a sensibility "in which a society exists because it is not possible to do without it but, at the same time, there is no longer a society" (Michaud 1980, p. 186). Under the monopoly of usefulness and the obsession with security as a guest, the social bond ends up deteriorating, curtailed in the format of a mere contractual interdependence. Michaud explains how the progress of modernity caused a *décalage* in the normative sphere by removing the rank of absoluteness linked to the institutional frameworks with consensual weight. This debilitated the chances for the emergence of a normative-referential criterion that could settle the unrest which arises inside every human group with the aim of getting self-vaccinated before an internal fracture resulting from the proliferation of revenges dictated by the random will of each individual. The toll is a caricatured collective conscience.

Late modernity deepens the blurring of the "fixed point" as a coordinating axis for the social sphere, thus favoring the tendency to a diverse, subjective, and variable assessment of every event. The radical relativization of the "fixed point" results in powerlessness to keep internal violence at bay, a situation favored by the fact that each social actor feels that they have a preferential right to claim the status of "fixed point" for their world interpretation scheme with no other legitimacy than the one based on belief or desire, i.e., "violence is the unilaterality in the affirmation of an individual or of a group in the social sphere conceived as playing games with those unilateral attitudes; in other words, an economy of violence; or expressed differently, a world without rules, without stability, and eventually without any predictability whatsoever" (Michaud 1980, p. 185). This final landscape results from a dis-involvement in relation to the moral sacrifice demanded by the "fixed point". After the institutionalism of normative rules has disappeared, the hostility between wills turns into fate, into an orgy of enmities with neither measure nor purpose reoriented towards the very heart of the collectivity. This serves to temporarily dispel the situations of structural unease emanated from a misunderstanding that contaminates social relationships, concealing what we do not want to see—the dissolution of the social sphere.[20]

Nonetheless, since the tragical thing is the shadow battered in the circle of the political prophylaxis urged by modernity, seeking to offset this, late modern societies focus on maximizing security, on indefatigably striving to reach a "zero risk" situation, and on removing antagonism at any rate, thus legitimizing themselves in accordance with the motto of a "perfection-oriented fanaticism" (Baudry 1986, p. 12), in short, demoralizing conflict, putting it on a level with a germ of harmful violence, and all of this weighted by the silencing of a violence that is endogenous to the actual social system—the one obsessed with reaching at all costs the fiction of a complete pacification of the social context (Baudry 1986, pp. 11–15; 2004, pp. 25–66). Once the conflict has been deprived of rituality formulas, a deadly de-vitalization of the social sphere is sown (Baudry 1986, p. 13), quarantining an "irrepressible desire to live" which is synonymous with creative and renewing disorder (Maffesoli 1984, p. 12) and seeing how a sordid floating violence—permanently about to break out—keeps growing in the ambiance. The typology of sacrificial victimization is consequently framed within an— even sometimes excessive—emphasis

20 This de-ritualized and democratized sacrificial practice, which comes as a result of negating the social aspects, coincides with a desire to embrace the vertigo of anomie. It does not matter which expression is used: "external phenomena" harboring a "radical alterity" under a motive to reencounter with "the cursed part" (Bataille) exorcized without a homeostatic replacement in modern culture (Baudrillard 1991, pp. 123 and ff.); "the obscene" as an exhibition of excrescence at the height of indifference to modern values (Baudrillard 1984, pp. 51–73); "wild sacred" as a rebound when faced with an instituted world where the sacred prevails in a tamed format (Bastide 1997, pp. 209 and ff.) or the violence that goes against *political rationality* (Carretero 2009). This anomic expressiveness would be betraying something paradoxical: "at the same time, loss of meaning and construction of meaning; de-subjetivation, but also subjetivation: (Wieviorka 2001, p. 346). One could state that the nihilist experience around the premises of modernity exacerbates variations of anomic meaning inextricably linked to that modernity.

on negating the social context, eradicating any possible ritualism able to forge a collective communion.[21]

3.2.3. New Liturgical Representation of Sacrificial Victimization

In late modern societies, the media-virtual universe is the stage which serves to project archetypal anthropological structures that serve as a resonance box for the ghosts that inhabit the collective imaginary.[22] A new public square where the everyday magma is "hypervisualized" (Imbert 2001) and where the aforesaid universe simultaneously pervades it. The ritual element is not immune to the rules of the game which apply to media-digital representation, being absorbed by its scenography, at the cost of being stripped to some extent of its solemnity. With it, the idiosyncrasy of sacrificial victimization adapts to the liturgical format of mass culture codes.

We can firstly deal with the fragility of the established meaning where daily life bathes (Schütz 1962; Berger and Luckmann 1986). Heidegger argued that everydayness relies upon a frail film made up of "gossip talk"—an anonymous, unfounded language which permeates it, a circulating speech whose authority rests upon repeating something with no support other than the fact of having been said before and the only purpose of which consists in extending a packaged interpretation of the world within reach of anybody. The everyday world is supported on a "gossip talk" which contaminates both how the world is seen and the way in which it must be seen (Heidegger 2000, pp. 186–200). The German thinker refers to this "flattening" of possibilities under the domination of a "One" using the word "publicity", a reading of the world which comprises what is taken for granted and what is accessible to everybody. A world of meanings built from "gossip talk" is anchored in insecure certainties, which makes it easier for tittle-tattle to appropriate everydayness. Elias and Scotson (2016, pp. 168–85) show how that world reinforces cohesion between individuals and how its effectiveness lies in being a twisted half-truth. It can be described as an a-structural weapon to strengthen social control, delving deeper into the disruption that a specific behavior represents with respect to conventional standards. Gossip unifies behaviors pursuant to prejudices which are reluctant to problematize arguments. Something similar takes place in the gossip talk converted into a micro-sociological category, a way of "speaking idly for talk's sake", enjoying what one does not have and letting oneself be dragged by the fascination of alterity, being captivated by the life of the other (Imbert 1992, pp. 118–225). The reign of "idle talk" additionally harbors rumor, the energy of which consists in believing what we want to believe a priori, i.e., in a peculiar willingness to believe, and whose operability rests upon a deregulated management of information through informal spaces (Morin 1969). As a common denominator, tittle-tattle, gossip, or rumor are communicative resources of unproductive economy (Imbert 1992, p. 119) which make possible the emergence of a common synergy around something within a situation of semi-opacity. The secret assumption of that something allows a variety of individuals to meet and ultimately become interwoven in a mutual tuning of affections. To this is added a sociability lazily based on habit and custom, including feast, chat, bland conversation, verbosity, or friendship computer networks, testimonies of a "subterranean centrality"

21 The de-institutionalization of sacrificial rituality leads to a widespread expansion of self-sacrificial behaviors where health is put to the test. Le Breton (2018) has shown a range of sacrificial practices amongst young generations which would have eliminated the role foreseen by *les rites de passage* to the adult universe. These generations recover a trace of sacrality deprived of collective substance that tries to transgress the instituted *numens* with no material benefit in return, the mark left by the primitive sacrificial act surviving, where what has been sacrificed encourages a transformation of the subject's interiority self with a view to build and reconstruct his identity, a youthful identity parasitized by an unease which refers us back to the access to an excessive normalizing zeal of the adult world without relinquishing the intensity of young life, added to the non-existence of ways established for its channeling, which in turn induces the emergence of a personalized sacrificial rituality that feeds a paroxysm of risk and physical self-harm. Young people would be shaping a particular identity meaning stamped as a mark through a wound on the skin or succumbing to damaging behaviors (Le Breton 2017, pp. 37–82).

22 In this direction, see Morin (1966), Balandier (1994), Imbert (2002), Carretero (2007), Martínez-Lucena and Barraycoa (2012), and Martínez-Lucena et al. (2019).

which, in addition to being inertial and surreptitious, can hardly be integrated into rigid institutional frameworks (Maffesoli 1990, pp. 52–67).[23]

We find ourselves before an everyday opaque *locus* from the epistemic thresholds of the social pact, rarely metabolized from formal control devices. The consensus link is grounded on a systemic–structural domain of a political–legal nature, albeit simultaneously in this informal and a-institutional *locus* rooted in everydayness. Simmel expressed it as follows: "Men look at one another, are jealous of one another, write letters to one another, eat together, are kind or unkind to one another, apart from any visible interest; the gratitude generated by an altruistic service has the power of an unbreakable bond, a man asks another for directions, men get dressed and do themselves up for one another, and all of these as well as a thousand other momentary or lasting, conscious or unconscious, ephemeral or fertile relationships, which take place between one individual and another, and from which we have arbitrarily singled out these examples, mutually link us at all times" (Simmel 1986, pp. 29–30). It is in this daily *locus* thrown onto the media-digital resonance box, where sociability is done and undone within an endless variety of "microscopic-molecular processes" (Simmel 1986), that late modernity essentially relocates the sacrificial phenomenology inherent to every culture in a self-backfed, restless, ending return trip from everydayness to the media-digital context. The forces of evil, at other times contained as elements violating the normative order, are not averted but are exhibited without any restrictions. This allows for the catharsis of endogenous violence (Girard) to be recycled under the form of a publicized catharsis released onto the media scene (Imbert 1992) and later onto the digital one. The outcome is a democratized face of the sacrificial praxis. Thus, mass media, social networking sites, or TV sets appear as platforms which project a sacrificial ghost in search of scapegoats. In late modern culture, nobody is expressly assigned the role of *sacrificer* and, at the same time, that role is assigned to everybody, the spirit of sacrificial victimization being stripped of any halo of sacrality.[24]

This is attested by the wide range of conspiracies, accusations, or stigmatizations encouraged by a merely phobic motivation. The rise of expressiveness correlates with the decline in the authority of institutional devices and, in parallel, with the inclination to a coexistence that is slave to the whims of subjectivism. Nobody would be excluded from this, since anybody could be half-judged by anybody else. The novelty of this phenomenon resides in its latent or low-intensity nature. Within this imaginary profile of victimization, despite the awareness of its wear and tear, the *numens* of modernity continue to be perceived as inviolable and obeyed without any reservations. The only problem lies in the fact that the enemy to be fought is not a dissidence cowering in difference, a situation systemically subsumed without any major difficulty in terms of a "different identity". On the contrary, it is a shadow of identity ambiguity which encumbers the transparency of the communicational flow on which sociability depends. If geometry arises as the archetype of the modern mind (Bauman 1995, p. 91), that aspect which resists being categorized upsets the confidence and instills uncertainty and insecurity within non-regulated interaction spaces. Its consummation is an abundance of attitudes governed by intolerance when not by a crusade spirit, a globalized village spirit where anybody can *in potential* become a sacrificial victim, the target of an indefinite victimization which hovers around. A ghost of distrust, in an osmosis between the media-digital sphere and everydayness, is fixed upon individuals whose main source of risk is the entanglement in a coexistence mediated by distancing, the one in which no "dense sociability" exists (Bauman 1995, p. 103), where everybody is a foreigner in everybody else's eyes. A sacrificial dimension is preserved—the aspiration to reach an expiation goal which acquits the collectivity within a purpose of

[23] "Why are people more interested in the gossip talk about Juan Guerra than in the speeches of political leaders? Because the former produces social cohesion and the latter do not: mass media are not only reservoirs of gossip, supermarkets which nourish tittle-tattle (in today's "global village", gossip talk about "celebrities" of art or politics plays the same role that gossip talk about the priest or the pharmacist used to fulfil in the old "global village") (Ibáñez 1990, pp. 18–19).

[24] Hanging like a sword of Damocles over any individual and favoring a "neigh borcracy", the neighborhood status which, assuming the control role, prevails over the citizenship status (Rodríguez Alzueta 2019).

catharsis which can put an end to the hostilities that cast a shadow over it day in day out, frightening away evil and trying to make the good reign attempt to restore a communion of consciences, more apparent than effective, dramatized on the media-digital universe.

4. Conclusions

According to the *numens* enhanced by modernity with the aim of consolidating a collective communion, emphasis has been placed on the persistence of an anthropological structure which, linked to sacrificial victimization, would have materialized in the former as a reclusion strategy. We elucidated the metamorphosis operated in the sacrificial imaginary of late modernity which is exposed to a de-institutionalized and a de-regulated ritual physiognomy inserted in a control logic consummated in an atmosphere of victimization channeled by social relationships which are favored by the complicity of a media-digital feedback. Evidence was provided not only to know why the characterization of this victimization constitutes a contrived attempt to achieve expiation—since it is not based on any real communion whatsoever—but also why, for this same reason, it would be doomed to failure in this task. Attention was equally paid to the fact that the reason for this failure can be found in the absence of commonly endorsed rites, in keeping with the exacerbation of a state that denies the social sphere and causes substantial damages in the collective link, and with a particular set of clearly unsuccessful strategies implemented for the purpose of filling the void in the former.

Funding: This research received no external funding.

Conflicts of Interest: The author declares no conflict of interest.

References

Agamben, Giorgio. 1998. *Homo sacer. El poder soberano y la nuda vida*. Valencia: Pre-Textos.
Anderson, Benedict. 1993. *Comunidades imaginadas. Reflexiones sobre el origen y la difusión del nacionalismo*. México City: FCE.
Arendt, Hannah. 2012. *Eichmann en Jerusalén*. Barcelona: Lumen.
Balandier, Georges. 1994. *El poder en escenas. De la representación del poder al poder de la representación*. Barcelona: Paidós.
Balandier, Georges. 1996. *El desorden. La teoría del caos y las ciencias sociales*. Barcelona: Gedisa.
Bastide, Roger. 1997. *Le sacré sauvage*. Paris: Stock.
Bataille, Georges. 1987. *La parte maldita*. Barcelona: Icaria.
Bataille, Georges. 1993. *El Estado y el problema del fascismo*. Valencia: Pretextos.
Baudrillard, Jean. 1980. *El intercambio simbólico y la muerte*. Caracas: Monte Avila.
Baudrillard, Jean. 1984. *Las estrategias fatales*. Barcelona: Anagrama.
Baudrillard, Jean. 1991. *La transparencia del mal*. Barcelona: Anagrama.
Baudrillard, Jean. 1996. *El espejo de la producción*. Barcelona: Gedisa.
Baudry, Patrick. 1986. *Une sociologie du tragique*. Paris: Cerf.
Baudry, Patrick. 2004. *Violences Invisibles. Corps, Monde Urbain, Singularité*. Paris: Passant.
Bauman, Zygmunt. 1995. Modernidad y ambivalencia. In *Las consecuencias perversas de la modernidad*. Barcelona: Anthropos, pp. 73–120.
Bell, Daniel. 1987. *Las contradicciones culturales del capitalismo*. Madrid: Alianza.
Bellah, Robert. 1967. Religion in America. *Dædalus, Journal of the American Academy of Arts and Sciences* 96: 1–21.
Bellah, Robert, Richard Madsen, William M. Sullivan, Ann Swidler, and Steven M. Tipton. 1989. *Hábitos del corazón*. Madrid: Alianza.
Benjamin, Walter. 1998. *Para una crítica de la violencia y otrosensayos. Iluminaciones IV*. Madrid: Taurus.
Berger, Peter L., and Thomas Luckmann. 1986. *La construcción social de la realidad*. Buenos Aires: Amorrortu.
Bergesen, Albert J. 1978. A Durkheimian Theory of "Witch-Hunts" with the Chinese Cultural Revolution of 1966–1969 as an Example. *Journal for the Scientific Study of Religion* 17: 19–29. [CrossRef]
Bergua, José Á. 2007. *Lo social instituyente. Materiales para una sociología no clásica*. Zaragoza: Prensas Universitarias de Zaragoza.
Beriain, Josetxo. 2007. Las máscaras de la violencia colectiva: Chivo expiatorio-mártir, héroe nacional y suicida-bomba. *Sociológica* 22: 77–118.
Beriain, Josetxo. 2009. Mártir y suicida: Dos caras del terror sacrificial moderno. *Anthropos* 222: 23–45.
Beriain, Josetxo. 2017. Las metamorfosis del don: Ofrenda, sacrificio, gracia, substitute técnico de Dios y vida regalada. *Política y Sociedad* 54: 645–67.
Bury, John. 2009. *La idea de progreso*. Madrid: Alianza.
Caillé, Alain. 2000. *Anthropologie du don*. Paris: La Découverte.
Callois, Roger. 1996. *El hombre y lo sagrado*. México City: FCE.
Canetti, Elias. 1983. *Masa y poder*. Madrid: Alianza.

Caro Baroja, Julio. 1970. *Inquisición, brujería y criptojudaísmo*. Barcelona: Ariel.
Carretero, A. Enrique. 2003. Religiosidades intersticiales. La metamorfosis de lo sagrado en las sociedades actuales. *Gazeta de Antropología* 19: 24. [CrossRef]
Carretero, A. Enrique. 2007. Lo mediático y lo social. Una compleja interacción. In *Proyectar Imaginarios*. Edited by Pardo Neyla. Bogotá: Instituto de Estudios en Comunicación y Cultura/Sociedad Cultural La Balsa, pp. 102–32.
Carretero, A. Enrique. 2009. Imaginario y violencia intracomunitaria. La racionalidad política y las formas anómicas de presentación de la violencia en las sociedades posmodernas. *Praxis Sociológica* 13: 38–67.
Carretero, A. Enrique. 2016. El imaginario social en torno al mal en la cultura posmoderna: De un pathos normativamente contenido a un pathos mediáticamente consentido. *Imagonautas, Revista Interdisciplinaria sobre ImaginariosSociales* 7: 1–20.
Casquete, Jesús. 2007. Religiones políticas y héroes patrios. *Papers* 84: 129–38. [CrossRef]
Cerruti, Pedro. 2010. El sacrificio como matriz jurídico-política: Crítica del fundamento biopolítico de la comunidad. *Pléyade* 3: 227–45.
Clastres, Pierre. 2004. *Arqueología de la violencia: La Guerra en las sociedades primitivas*. México City: FCE.
De Maistre, Joseph. 2019. Elucidación sobre los sacrificios. *Stylos*, pp. 106–282. Available online: https://repositorio.uca.edu.ar/bitstream/123456789/9698/1/elucidacion-sacrificios-maistre.pdf (accessed on 2 August 2020).
Deleuze, Gilles. 1995. *Conversaciones*. Valencia: Pre-Textos.
Delgado, Manuel. 1999. *El animal público*. Barcelona: Anagrama.
Delgado, Manuel. 2001. *Luces iconoclastas: Anticlericalismo, espacio y ritual en la España contemporánea*. Barcelona: Ariel.
Delgado, Manuel. 2008. Las instituciones atroces. Turbas rituales y violencia iconoclasta en la España contemporánea. In *Religión y violencia*. Edited by Lanceros Patxi and Díez de Velasco Francisco. Madrid: Círculo de Bellas Artes, pp. 181–211.
Delgado, Manuel, and Sarai Martín López. 2019. La violencia contra lo sagrado. Profanación y sacrilegio: Una tipología. *Vínculos de Historia* 8: 171–88. [CrossRef]
Díez de Velasco, Francisco. 2008. Sentidos, violencias, religiones. In *Religión y violencia*. Edited by Patxi Lanceros and Francisco Díez de Velasco. Madrid: Círculo de Bellas Artes, pp. 251–86.
Dodds, Eric R. 2006. *Los griegos y lo irracional*. Madrid: Alianza.
Douglas, Mary. 1973. *Pureza y peligro: Un análisis de los conceptos de contaminación y tabú*. Madrid: Siglo XXI.
Douglas, Mary. 1978. *Símbolos naturales*. Madrid: Alianza.
Durand, Gilbert. 1982. *Las estructuras antropológicas de lo imaginario*. Madrid: Taurus.
Durand, Gilbert. 1996. *Introduction à la mythodologie. Mythes et sociétés*. Paris: Albin Michel.
Durkheim, Émile. 1982. *Las formas elementales de la vida religiosa*. Madrid: Akal.
Duvignaud, Jean. 1979. *El sacrificio imposible*. México City: FCE.
Duvignaud, Jean. 1990. *Herejía y subversión*. Barcelona: Icaria.
Elias, Norbert. 1989. *El proceso de civilización. Investigaciones sociogenéticas y psicogenéticas*. México City: FCE.
Elias, Norbert, and John L. Scotson. 2016. *Establecidos y marginados. Una investigación sociológica sobre problemas comunitarios*. México City: FCE.
Esposito, Roberto. 2003. *Communitas. Origen y destino de la comunidad*. Buenos Aires: Amorrortu.
Estruch, Joan. 1994. El mito de la secularización. In *Formasmodernas de religión*. Edited by Rafael Díaz Salázar, Salvador Giner and Fernando Velasco. Madrid: Alianza, pp. 266–80.
Evans-Pritchard, Edward E. 1977. *Los nuer*. Barcelona: Anagrama.
Ferro, Mariño. 1984. *Satán, sus siervas las brujas y la religión del mal*. Vigo: Xerais.
Foucault, Michel. 1967. *Historia de la locura en la época clásica*. México City: FCE, vol. 1.
Foucault, Michel. 1992. Verdad y poder. Entrevista con Fontana, M. In *Microfísica del poder*. Madrid: La Piqueta, pp. 175–89.
Foucault, Michel. 1994. *Vigilar y castigar*. Madrid: Siglo XXI.
Foucault, Michel. 2006. *Seguridad, territorio, población*. Buenos Aires: FCE.
Frazer, James G. 1981. *La rama dorada*. México City: FCE.
Freud, Sigmund. 1988. *Tótem y tabú*. Madrid: Alianza.
Girard, René. 1983. *La violencia y lo sagrado*. Barcelona: Anagrama.
Girard, René. 2012. *El sacrificio*. Madrid: Encuentro.
Godbout, Jacques T., and Alain Caillé. 2000. *L'esprit du don*. Paris: La Découverte.
Goffman, Ervin. 1972. *Internados. Ensayo sobre la situación social de los enfermos mentales*. Buenos Aires: Amorrortu.
Goffman, Ervin. 1981. *La presentación de la persona en la vida cotidiana*. Buenos Aires: Amorrortu.
Gutiérrez Martínez, Daniel. 2010. De las conceptualizaciones de las religiones a las concepciones de las creencias: A manera de introducción. In *Religiosidades y creencias. Diversidades de lo simbólico en el mundo actual*. México City: Zinacantepec/Colegio Mexiquense, pp. 9–44.
Heidegger, Martin. 2000. *El ser y el tiempo*. México City: FCE.
Hénaff, Marcel. 2002. *Le prix de la verité. Le don, l'argent, la philosophie*. Paris: Seuil.
Ibáñez, Jesús. 1990. Prólogo. In *El tiempo de las tribus*. Barcelona: Icaria, pp. 8–19.
Imbert, Gerard. 1992. *Los escenarios de la violencia. Conductas anómicas y orden social en la España actual*. Barcelona: Icaria.

Imbert, Gerard. 2001. La hipervisibilidad televisiva: Nuevos imaginarios/nuevosrituales comunicativos In *Revista electrónica. Instituto de Cultura y Tecnología*. Madrid: Universidad Carlos III. Available online: https://www.um.es/tic/LECTURAS%20FCI-II/FCI-II%20tema2textocomplementario2.pdf (accessed on 4 September 2020).

Imbert, G. 2002. Azar, conflicto, accidente, catástrofe: Figures arcaicas en el discurso posmoderno. *Trama y fondo: Revista de cultura* 12: 19–30.

Jiménez Lozano, José. 1999. Me aterra lo sagrado. *Archipiélago. Cuadernos de crítica de la cultura* 36: 11–15.

Juan, Salvador. 2013. *La Escuela francesa de socioantropología*. Valencia: Universitat de València. Servei de Publicacions.

Jung, Carl G. 1966. *El hombre y sus símbolos*. Madrid: Aguilar.

Le Breton, David. 2017. *El cuerpo herido. Identidades estalladas contemporáneas*. Buenos Aires: Topía.

Le Breton, David. 2018. *La piel y la huella*. México City: Paradiso.

Lévi-Strauss, Claude. 1964. *El pensamiento salvaje*. México City: FCE.

Lorio, Natalia. 2013. La potencia de lo sagrado en la comunidad. Un rastreo de Durkheim a Bataille en el Colegio de sociología. *ARETÉ. Revista de Filosofía* XXV: 111–31.

Luhmann, Niklas. 1996. *Confianza*. Barcelona: Anthropos.

Luhmann, Niklas. 1998. *Complejidad y modernidad: De la unidad a la diferencia*. Translated and Edited by Josetxo Beriain, and Blanco J. M. García. Madrid: Trotta.

Luhmann, Niklas. 2007. *La religion de la sociedad*. Madrid: Trotta.

Maffesoli, Michel. 1984. *Essai sur la violence banale et fondatrice*. Paris: Méridiens/Klincksieck.

Maffesoli, Michel. 1990. *El tiempo de las tribus. El declive del individualism en las sociedades de masa*. Barcelona: Icaria.

Maffesoli, Michel. 1993. *El conocimiento ordinario. Compendio de sociología*. México City: FCE.

Maffesoli, Michel. 2002. *La part du diable*. Paris: Flammarion.

Maffesoli, Michel. 2005. *La transfiguración de lo político. La tribalización del mundo posmoderno*. Barcelona: Herder.

Magdalena, Nélida A. 2015. Los oscuros dioses en una época sin rituales sacrificiales. La religiosidad del neurotico en transferencia. In *VII Congreso Internacional de Investigación y Práctica Profesional en Psicología XXII. Jornadas de Investigación XI Encuentro de Investigadores en Psicología del MERCOSUR*. Buenos Aires: Facultad de Psicología, Universidad de Buenos Aires, pp. 421–23. Available online: https://www.aacademica.org/000-015/794.pdf (accessed on 20 August 2020).

Martínez-Lucena, Jorge, and Javier Barraycoa. 2012. El zombi y el totalitarismo: De Hannah Arendt a la teoría de los imaginarios. *Imagonautas: Revista Interdisciplinaria sobre Imaginarios Sociales* 2: 97–118.

Martínez-Lucena, Jorge, González de León Berini, and Arturo Abbate Stefano, eds. 2019. *Control social e imaginarios en las teleseries actuales*. Barcelona: UOC.

Mauss, Marcel, and Henri Hubert. 1970. Ensayo sobre la naturaleza y función del sacrificio. In *Obras*. Barcelona: Barral.

Michaud, Yves. 1980. *Violencia y política*. Paris: Ruedo Ibérico.

Morin, Edgard. 1966. *El espíritu del tiempo*. Madrid: Taurus.

Morin, Edgard. 1969. *La rumeur d 'Orléans*. Paris: Seuil.

Páez, Laura, ed. 2002. *La Escuela Francesa de Sociología. Ensayos y textos*. México City: UNAM.

Polanyi, Karl. 1997. *La gran transformación. Crítica del liberalism económico*. Madrid: La Piqueta.

Prades, José A. 1998. *Lo sagrado. Del mundo arcaico a la modernidad*. Barcelona: Península.

Robertson-Smith, William. 1894. *Lectures on the religion of the semites*. London: Adam and Charles Black.

Rodríguez Alzueta, Esteban. 2019. *Vecinocracia. Olfato social y linchamientos*. Buenos Aires: Eme.

Rusche, Georg, and Otto Kirchheimer. 1984. *Pena y estructura social*. Bogotá: Temis.

Sánchez Capdequí, Celso. 1998. Las formas de la religion en la sociedadmoderna. *Papers* 54: 159–85.

Sánchez Capdequí, Celso, and A. Enrique Carretero. 2006. Le fondement de l'experience religieuse et sa versatilité dans le monde contemporaine. *Sociétés. Revue des Sciences humaines et Sociales* 93: 103–17.

Schütz, Alfred. 1962. *El problema de la realidad social*. Buenos Aires: Amorrortu.

Simmel, Georg. 1986. *Sociología 1. Estudio sobre las formas de socialización*. Madrid: Alianza.

Simmel, Georg. 2011. *El pobre*. Madrid: Sequitur.

Turner, Victor. 1980. *La selva de los símbolos. Aspectos del ritual ndembu*. Madrid: Taurus.

Turner, Victor. 1988. *El proceso ritual. Estructura y antiestructura*. Madrid: Taurus.

Weber, Max. 1993. *Economía y sociedad*. México City: FCE.

Wieviorka, Michel. 2001. La violencia: Destrucción y constitución del sujeto. *Espacio abierto* 10: 337–47.

Article

A Maximal Understanding of Sacrifice: Bataille, Richard Wagner, Pilgrimage and the Bayreuth Festival

Philip Smith [1,*] and Florian Stoll [2,*]

1 Department of Sociology, Yale University, New Haven, CT 06511, USA
2 Research Center Global Dynamics (Recent Globe), Leipzig University, Strohsackpassage, Nikolaistr. 6-10, 04109 Leipzig, Germany
* Correspondence: philip.smith@yale.edu (P.S.); florian.stoll@uni-leipzig.de (F.S.)

Abstract: This paper calls for a broad conception of sacrifice to be developed as a resource for cultural sociology. It argues the term was framed too narrowly in the classical work of Hubert and Mauss. The later approach of Bataille permits a maximal understanding of sacrifice as non-utilitarian expenditures of money, energy, passion and effort directed towards the experience of transcendence. From this perspective, pilgrimage can be understood as a specific modality of sacrificial activity. This paper applies this understanding of sacrifice and pilgrimage to the annual Bayreuth "Wagner" Festival in Germany. Drawing on a multi-year mixed-methods study involving ethnography, semi-structured interviews and historical research, the article traces sacrificial expenditures at the level of individual festival attendees. These include financial costs, arduous travel, dedicated research of the artworks, and disciplines of the body. Some are lucky enough to experience transcendence in the form of deep emotional experience, and a sense of contact with sacred spaces and forces. Our study is intended as an exemplary paradigm case that can be drawn upon analogically by scholars. We suggest that other aspects of social experience, including many that are more 'everyday', can be understood through a maximal model of sacrifice and that a rigorous, wider comparative sociology could be developed using this tool.

Keywords: sacrifice; pilgrimage; sacred; festivals; Wagner; Bayreuth; Durkheim; opera

Citation: Smith, Philip, and Florian Stoll. 2021. A Maximal Understanding of Sacrifice: Bataille, Richard Wagner, Pilgrimage and the Bayreuth Festival. *Religions* 12: 48. https://doi.org/10.3390/rel12010048

Received: 20 December 2020
Accepted: 7 January 2021
Published: 11 January 2021

Narrowly understood sacrifice conjures the image of priests, shamans, altars, dead animals and burnt offerings—in short religious contexts and certain forms of institutionalized ritual. However, in this paper, we argue that a fluid and 'maximal' understanding of sacrifice is more useful for social science. Sacrifice can be understood as those irrational, evanescent expenditures—of money, time, energy, goods, concern, passion, attention, discipline—in life that generate the experience of a more meaningful personal and collective existence, and, at the limit, an encounter with the sacred. Of course, the experiences and their attendant sacrifices will be of varying intensity and take diverse forms in particular lifeworlds, settings and social contexts. These require specification, and such is the task for the sociologist seeking to make the general model tractable as a way of seeing. We illustrate all this with reference to the composer Richard Wagner's Bayreuth Festival, a major cultural event that takes place every summer in Germany. We demonstrate that a maximal theory of sacrifice helps us understand this particular gathering, along with its participant motivations, experiences and patterns of action. We also show that it belongs to the specific sacrifice modality of 'pilgrimage'. This has its own dynamics and needs to be understood from within as a form of life. To make sense of our approach and what it contributes, we need first to detour into the history of cultural theory.

Within the Durkheimian tradition, the canonical resource for understanding sacrifice is the one-hundred-and-nine-page essay by Henri Hubert and Marcel Mauss, first published in the *Année Sociologique* in 1899. It had taken several months to write—longer than expected despite the persistent urgings and micro-management of Émile Durkheim himself. Under the auspices of E.E. Evans-Pritchard, the item was eventually translated into English as many as six decades later and published as a small book (Hubert and Mauss 1964). The venture was driven more by intellectual devotion than by popular demand in the academic community. The Foreword by Evans-Pritchard notes and accounts for neglect: "If little reference to this Essay has been made in recent decades it is perhaps due to a lack of interest among sociologists and anthropologists in religion and therefore its most fundamental rite" (Evans-Pritchard 1964, p. viii).

Does this statement hold true today? In one way, no. Since the late-1980s, cultural sociology, to take just one intellectual field, has drawn heavily on the *Elementary Forms of Religious Life* (Durkheim 1995) to develop a comprehensive understanding not so much of religion but rather of the wider 'religious' dimensions of society (Smith 2020). According to the Strong Program, for example, the world is classified into the sacred and profane (Alexander and Smith 2010); and for Collins (2004), social structure consists of nested and contending interaction rituals that generate totemic power. But if the religious/ritual quality of social life has moved center stage, it is not clear that Hubert and Mauss's essay has garnered that much interest as a consequence. At the time of writing, there are a little over 2000 citations listed on Google Scholar for all versions of this 'classic' and, following through, we find that many of those are in the history of theory mode. The short book is often noted but less frequently used. Why so?

At the heart of the limited appeal is the narrow understanding in *Essai sur la Nature et la Fonction du Sacrifice*. 'Sacrifice' here is not a particularly useful, supple or general concept. For all their pioneering brilliance, Hubert and Mauss were still attempting to intervene in the anthropological debates of the late-Victorian era that had been set in motion by Tylor and Frazer, and above all Robertson Smith. In his *Religion of the Semites*, Robertson Smith (2002) argued that the origins of sacrifice lay in totemism, and notably the collective meal during which the tribe generated a sense of common identity through consumption of the totemic animal. With the emergence of agriculture, the domestication of animals and increases in social complexity, the feast had been replaced with an expiatory sacrifice that was mediated, symbolic or tokenistic in form and guided by priestly specialists. True enough, Hubert and Mauss fault Robertson Smith's method and his speculative efforts at a reconstruction of the evolution of a complex social and cultural institution. And to their credit, in its place, they offer a more universal or structural vision of the necessary and functional elements of a sacrificial system. Yet as they trawl through examples from Vedic, Christian, Greek, Babylonian and Egyptian civilizations in a virtuoso display of ethnological encyclopedism, the image remains one of priests and religious devotees killing and destroying upon altars in order to placate Gods or bring about desired social ends. Given the diversity of uses to which the word 'sacrifice' is applied in everyday life, and the later evolution of Durkheimian theory itself, this is a curiously literal/liturgical understanding that does not provide the concept with room to breathe.

Nearly three decades on from his collaboration with Hubert, Marcel Mauss was a more ambitious and fluid thinker. In his celebrated essay on *The Gift* from 1925, he mapped out a landscape of generalized symbolic flows and non-accumulative exchanges as pivotal to the moral economy of primitive society (Mauss 1954). Whereas modern capitalism was marked in his view by utilitarian calculation, commodities, and profit seeking, these gift economies involved generosity, reciprocity, and the sharing and destruction of surplus goods. Pivotal to Mauss arriving at this point of insight was an impactful reading of the work of Bronislaw Malinowski. In his masterwork *Argonauts of the Western Pacific*, first published three years before *The Gift* in 1922, Malinowski (1932) offered a vivid description and analysis of the *kula* ring. This saw venerated ceremonial objects, such as shell jewelry, circulate around remote islands in a process that involved risky open canoe voyages over hundreds of kilometers, followed by elaborate ceremonies of welcome and exchange. The recipients of the artefacts were obligated to pass them on. They could not hoard, nor sell the valuable items for profit. Argonauts emphasized that *kula* exchange was clearly differentiated from the market exchange in the Trobriand Islands that was known as *gimwali*. This involved more conventional understandings of pricing, fungibility and commensurability.

Although *kula* exchange appeared irrational in terms of any risk/energy/reward calculus Malinowski, in his trademark effort to foreground primitive rationality, stressed the benefits to the individual of being a central player in *kula* exchange networks. In receiving and giving away goods, they gained social visibility and prestige. Mauss, ever the Durkheimian, differed in seeing *kula* exchange as an act that was more about collectivities than the desires of individuals and the politics of village life. It tied entire societies together, prevented war and was about solidarity and shared ritual experience.

So Mauss perceived a deep, binding sociality arising from the movement of sacred items around the *kula* ring of Melanesia and drew parallels with another constitutive social fact that had seized the attention of ethnologists—the wholesale destruction of trade goods in the potlatch of the tribes of the Pacific Northwest. Yet he never quite connected his thinking in *The Gift* to his earlier work with Hubert via an adequate discussion of how sacrifice could be seen as imminent to such aspects of ritualized, solidaristic economic-social activity. Symptomatic of this was the fact that whereas gift exchange was broadly understood as moral and communicative, potlatch was viewed with a hint of suspicion as a competitive activity—ironically a reading not unlike that which Malinowski had made of the *kula* ring. There was a certain asymmetry to the evaluation that prevented Mauss from seeing positive dimensions to destruction/consumption/waste as well as exchange and sharing. Furthermore, Mauss had folded in a critique of modernity into his description of the gift economy. His normatively loaded examples suggested that it mostly belonged in the simple and wise societies of the past. Mauss had painted himself into the ethnologist's corner. How, then, could the lessons from *The Gift* be deployed to understand modernity?

It was to be the maverick Durkheimian Bataille (1985, 1988) who joined the dots in the two decades following the publication of Mauss's essay. He did so in three important ways. First, he explicitly connected Mauss's work on primitive economics to the theme of sacrifice, suggesting that all operations of consumption have a sacrificial logic. They are the forms of destruction and expenditure that curiously enough hold society together. Secondly, he insisted that modern society also had the primitive economic operations Mauss had identified and that sacred forces were behind these. Whereas Mauss had ghosted in a critique of modernity, Bataille thought that we were not particularly modern. Thirdly, Bataille offers a more general resource and set of examples that drag the study of 'sacrifice' away from its theological origins and so expand the concept's explanatory reach to explain a raft of activities and passions. The consequence? Bataille was able to consider a large proportion of the modern political economy to be founded on logics of sacrifice.

Drawing on Mauss on gift economies, Durkheim (1995) on the ambivalence of the sacred and Hertz (1960) on the bilateral symbolism of left and right, as well as upon surrealism, Freud and a general distrust of Western civilization, Bataille developed a complex and somewhat cosmological vision of cultural life. Energy pours into the planet from the sun. This surplus energy leads to the production of life forms. Human cultures feed off these, developing complex symbolic systems and patterns of association. Cosmological balance can only be restored when that surplus is destroyed. In Bataille's vision, the core of sociality lay in these irrational expenditures of a sacrificial nature that enabled a flow and translation of energies. But he did not see 'sacrifice' in the narrow way of Robertson Smith, or Hubert and Mauss. Rather, Bataille opens up a landscape that includes acts of violence, eroticism and desire, individual efforts towards the experience of sacred terror, the love of luxury, and the sybaritic pleasures of indulgence and decadence. All these were essentially non-utilitarian, ceremonial activities that fruitfully disposed of surplus capital and energy rather than engaging in rational, profane accumulation. Through these expenditures, social life was valorized, completed and made whole. It is no accident that in his masterwork of the 1940s *The Accursed Share*, Bataille (1988) speaks admiringly and at length of Aztek human sacrifice as a paradigm case. He sees in the complicity of the victims and the extraordinary ceremonial courtesies extended towards them a refined and knowing gesture. The Aztecs, as he saw them, had a deep civilizational awareness of the productive and energizing forces attending to destructive expenditures of human life.

It is too bad that Bataille's message has never quite stuck, at least in sociology (Smith 2020). He writes in a style that is too literary and elusive, too far from positivism and too full of manifesto-like calls for anarchy. Sociologists in the academy have never quite known what to do with him. Yet the core themes have filtered through to some of the most significant social theory of our time (Riley 2013). Hence, Foucault has written on experiences of the extreme, Baudrillard on consumption rather than production as pivotal to modern capitalism, and Kristeva on the strange attractions of abjection and horror. Perhaps the most influential item attempting to use Bataille in a more conventional mode is Miller's (1998) much cited *Theory of Shopping*. Based on an ethnography of a north London high street, Miller sees shopping not as a practical activity aimed at provisioning the household but as a series of irrational expenditures that bind society together. It is a vehicle for expressing love, pampering the self, generating ritual solidarity and celebrating social life through a privileging of non-utilitarian, reckless choices that often go against our own better judgement. This is done as much by buying a new shade of lipstick as a pound of potatoes. With the sacrifice of money for that which is not really needed, with the treat, gift, or upsell come emotional and spiritual rewards that affirm and build human relationships.

But shopping is just one domain where the sacrifice/expenditure model holds. Consider the following grist for Bataille's all-purpose mill: Formula One motorsport; high fashion and fast fashion; New Year firework displays; Roller Derby; the bullfight; boxing; hobbies that soak up discretionary income; risky sex with strangers; gift giving in the family; getting drunk on Friday night; women's overpriced haircuts and hair styling; super yachts or the much smaller boat that stretches the family budget; the circus; carnival; meticulously white sneakers in poor neighborhoods; funerals; Las Vegas; collecting garden gnomes. As the saying goes, the possibilities are endless. And note that each of these domains of irrational expenditure and theaters of excess has its own rules of the game, its own emotional rewards, subject positions, performative scripts and cultural codes. So the general vision of sacrifice needs to be specified through detailed sociological inquiry into defined spheres of activity in particular lifeworlds. Bataille's model points to a general direction of intellectual travel, but it is just too imprecise to do the detailed explanatory mapping of empirical sociology. Rather, we need blueprints for the geometries of specific irrational expenditures. To illustrate the point, our paper applies the general thematic of sacrifice and expenditure to the specificities of Wagner fans at the Bayreuth Festival. It considers the search for aesthetic pleasure and the role of travel. In so doing, we make use of a developed intermediary resource in the cultural sociological toolbox—the concept of pilgrimage. This brings the gloss of 'sacrifice' down to earth in a middle range way. Now to our second detour into theory.

1. Pilgrimage

Within the Durkheimian tradition, the classical resource for understanding pilgrimage is far less well known than the Hubert and Mauss essay on sacrifice. Indeed, less than thirty years ago, Macclancy (1994, p. 32) noted in a review essay that he could find only one reference to it in the English literature and that it seemed unknown even to most pilgrimage specialists. The first English translation was not until the 1980s and even then it was not easy to find, buried he tells us in an edited collection. The item in question *Saint Besse: Etude d'un culte alpestre* dealt with pilgrimage to a shrine high in the Alps and was written by Durkheim's talented student Hertz (1913). Published in 1913, it tells how pilgrims came from two different Italian valleys to the shrine carrying a heavy statue, how there was an auction to raise funds, and how pilgrims chipped at rocks so as to have a relic to take home. Hertz recounts competition between cults from the two valleys and their divergent mythologies surrounding the Saint. The approach is somewhat ethnological, 'fact collecting' and theory light. In contrast to Hertz's work on classification and death, the wider lessons have to be extracted by readers. Namely, that in retrospect this is a pioneering study of the politics of ritual and contested collective memory.

It was, of course, Turner (1969, 1973) who much later put pilgrimage firmly on the map for cultural anthropology, sociology and to some extent also religious studies. He saw it as an institutionalized form of liminality. In the course of their journey away from routine social life with its hierarchies and fixed identities, pilgrims would experience an openness to the sacred, the deep solidarity known as 'communitas', and, with this, transformations of the self. Perhaps seeing Durkheim as a functionalist, Turner never recognized that many of his insights were prefigured in Durkheim's (1995) *Elementary Forms of Religious Life* but the parallels are clear (Smith 2020, p. 183). Importantly, while Hertz opens up the study of pilgrimage to themes of power, Turner comes closer to a more Geertzian understanding of the activity as expressive and meaningful. It exists precisely because it offers symbolic, social and emotional rewards. Like 'sacrifice', however, 'pilgrimage' was been hamstrung over the years by narrow applications to religious contexts—such as travel to Jerusalem, Mecca and saintly shrines in Spain, Latin America or India. It is through wider literatures such as tourism studies that we first started to see a broader range of activities through this lens. Australian backpackers travelling to the battlefields of Gallipoli in Turkey, for example, are encountering sacred national myths in a very personal, emotive and embodied way (West 2008). Their extended vacation is reasonably viewed as a pilgrimage, or at least of having pilgrimage-like properties during some phases of activity.

Finally, we arrive at the point in this paper where the concepts of 'pilgrimage' and 'sacrifice' can be united. Putting Turner into dialogue with Bataille, we see pilgrimage as a particular lifeworld form of non-rational, 'pointless' expenditure in pursuit of the sacred. Rather than engaging in productive labor and the accumulation of material goods, the pilgrim expends effort and capital as they enter a form of special time to engage in long distance travel, overcome trials and enter into contemplative activity that has no utilitarian purpose. Energy, physical and mental, is discharged freely in celebrating the sacred, deepening the self and generating existentially profound experiences. It is a sacrificial action of a serious kind. Notably pilgrimage requires effort, discipline, asceticism, abstinence and deferred gratification. Many of Bataille's favored illustrations of sacrificial logics involve easy pleasures—alcohol, orgies, luxury goods and bohemian decadence. Pilgrimage alerts us to the ways in which irrational expenditures can take a contrary form while being equally transient, non-utilitarian and non-accumulative. Suffering is a form of sacrifice.

2. Wagner, Bayreuth and Pilgrimage

This brings us to the 19th-century composer Richard Wagner and his deeply serious vision of art. Although increasingly indifferent to organized religion and an admirer of the atheist philosopher Schopenhauer, Wagner (1994) nonetheless had a spiritual understanding of social life. He saw art replacing religion as a provider of mythological and spiritual truths and as a source of solace. He also believed in the power of myth, the study of which could provide profound knowledge regarding the centrality of compassion, redemption, suffering, love and sin in meaningful human experience. His own artworks were conceived as bringing listeners closer to this set of understandings at both conscious and unconscious levels. It is no surprise, therefore, to see that themes of pilgrimage and sacrifice loom large in his oeuvre. Characters such as Lohengrin, Siegfried and Parsifal engage in journeys that have spiritual elements and result in trials, transformations, realizations and purifications of the self. As for sacrifice, this is shown to be central to the closure of rupture: Hans Sachs puts aside his desire and renounces his claims on Eva's affections so that the natural order of young love can be restored; Isolde dies so as to bring about a deeply spiritual erotic unity with Tristan and so reconcile 'day' and 'night'; Brünnhilde rejects the cursed *Ring des Nibelungen* and throws herself on Siegfried's funeral pyre to the sound of the *Erlösungs*-(resolution) *motif*. We cannot resist noting that these actions within the plot are absolutely consistent with Bataille's logic: sacrifice enables cosmological and social balance to be restored.

That said, we are not concerned in this paper with the mytho-poetics of Wagner's stage works but rather with his larger social project. In some ways, Wagner can be considered a spiritual social movement leader. He hoped that his final work, *Parsifal*, "would purify the world by bringing it to a state of Christian pity and renunciation" (Spotts 1994, p. 79). Drawing on the local tradition of community passion plays, he dubbed it a "sacred stage consecration play" (*Bühnenweihfestspiel*) and for a long time refused permission for it to be performed outside of Bayreuth. Despite the bourgeois tendencies in his personal lifestyle that have long fascinated and disappointed critics (Adorno 2005; Mann 1985), the composer and one-time revolutionary was always something of an egalitarian, ascetic purist when it came to the role of art in society. His theoretical and programmatic writings set out the stall. He despised the ways that opera was subordinated to ostentatious social display in Paris. The opera house he designed at Bayreuth was one in which every seat, deliberately uncomfortable, had a good view of the super-sized stage in a darkened auditorium. Tickets to the first Bayreuth Festival were available at low cost to subscribers who had demonstrated their loyalty to his project, rather than sold off to elites at high prices. His ambition was to develop a rather cult-like cooperative organization, led by himself, organized around the performance of his artworks. The remote location of Bayreuth was also something he praised. In contrast to big cities with their distractions, Bayreuth was—and is—a sleepy country town. Festival-goers would have to make a dedicated trip and, Wagner hoped, would be able to focus on his works seeking therein spiritual truths. There was something stripped back rather than hedonistic about the entire enterprise, despite the complexity and luxuriance of his music. Indeed, attending Bayreuth was conceived by Wagner as something of a pilgrimage: travel to an inconvenient and unfashionable destination, egalitarianism, pursuit of the sacred and self-knowledge were combined. If this was the intent, then what has been the experience?

Since 2015, we have conducted mixed-methods research on the Bayreuth Festival. Central to the project has been the collection of historical accounts, as well as a series of interviews with contemporary festival-goers and substantial ethnographic participation at the event. The accounts and interviews cover a range of topics including the experience of the festival and the town, responses to performances, thoughts about Bayreuth's troubled past due to associations with Hitler and antisemitism, and travel practicalities. Through an investigation of these resources, we can reconstruct the meanings of the event and the ways in which the sacrifice/pilgrimage model keys to this particular context. The themes: travel and movement to a liminal space, suffering and effort, learning and changing the self, egalitarian solidarity, transcendent experience, a sense of the sacred, an awareness of costs and of the activity as deeply meaningful. By attending to these, we come to understand the form of life that is involved in this particular sacrificial domain.

We note in starting that the language of pilgrimage is commonly applied by historians, critics and commentators to the journey to Bayreuth and to capture the mentality of attendees. It is an explicit part of the folk logic, an 'emic' element of Bayreuth-speak and not simply our 'etic' imposition from cultural theory. Hence, commentators make use of the pilgrimage theme to capture ascetic privation and liminality:

> *"Why content oneself with the annual ritual of the Bavarian Epidaurus, the very architecture of which was designed as an uncomfortable pilgrim's arena to make it easier for the seekers of total aesthetic phenomenon to leave behind the mean comforts of the common world and to raise their minds to the heights of art"* (Heller 1985, p. 19)

Or extremes of long-distance travel:

> *"Although there were fewer foreigners in the audiences during the early years of the century, Bayreuth retained its allure as a place of pilgrimage. The pilgrims came from as far away as China and California"* (Spotts 1994, p. 130)

Or self-transformation:

> *"Does this mean that the former "sanctuary" of the primarily German bourgeoisie is being transformed into an "adventure park" for a post-bourgeois high culture scene, a*

procession to the refuge of eternal truth for a "pilgrimage into the self"?" (Gebhardt and Zingerle 1998, p. 30; our translation)

Our focus here, though, is not so much on the interpretations of other intellectuals, which as we just saw can be laced with irony, and more on the folk logic of participants. How do those coming to Bayreuth view their experiences? What images and stories do they conjure when reflecting on their participation?

Our transcripts and the historical accounts we examined showed that talk about Bayreuth was surprisingly thick with talk about expenditures. These included money, cognitive mental energy and emotions. There is a pervasive sense of effort rather than ease. The efforts are combined with an underlying belief in the significance of the festival and its artworks. Seeing a performance in Bayreuth does not only demand financial spending but also focus, personal dedication and physical discipline. It is crucial to know the dramaturgy, the music and even the German lyrics well to have access to a deeper understanding of Wagner's works. There are no subtitles or translations provided during the performance. Listening to the several-hours-long operas, reading the libretto and even more mundane duties such as organizing tickets and selecting a tuxedo or a festive dress are part of a trip to Bayreuth. All these activities are more than simply 'fun'—they are in fact arduous.

Such expenditures make a visit to Bayreuth a very different experience than ordinary summer travels such as attending the Glastonbury Festival, doing a city trip to Rome, or visiting historical sites in Greece. From Bataille's perspective, of course, these also involve sacrificial 'pointless' expenditures. But with the *Festspiele* (festival), these expenses and efforts are more visible and intense. It is not a holiday or a cultural event—the logic is one of pilgrimage. And so going to the *Festspiele* involves a usually complicated trip to the inconveniently located German province, alongside mental and physical preparation to experience spiritual insight—similar to religious pilgrims who walk the Camino de Santiago or go to Mecca. Difficulties and inconveniences are an integral part of the festival. The investments force people to concentrate and so assist in the evolution of passion. In the logic of pilgrimage, the sacrifices involved bring meaning. And that meaning explains why people do not just do the 'sensible thing' and stay at home and watch the operas on DVD. Let us bring in some more detail and go through the sources of effort in a logical order.

3. Tickets

We begin with a difficulty that is hiding in plain sight: getting a ticket to a performance. For decades, the major share of tickets was given to clearly defined groups such as the Wagner supporters club *Freunde Bayreuths* (Friends of Bayreuth) with its expensive membership. Allocations were also made to core players in Germany's industrial and civil order, such as federal employees, trade unions and sponsoring corporations. Wagner's egalitarian dream had failed. The only regular way for opera fans to get tickets was to apply at the festival's dead letter office, then to move up the waiting list year by year. There was much rumor about how to game the inscrutable system. For example, the Munich-based *Süddeutsche Zeitung* published in 2008 a half-satirical, half-serious article how one can find alternatives to the estimated 10 years waiting time (Zinnecker 2008).

The improbable hunt for tickets demonstrates how the hope for a transcendent experience, waiting and suffering go hand in hand in this case. It is also something that takes the festival in multiple ways out of the exchange system of ordinary commodities and turns participation into something more ritualized and mysterious or providential. The tickets arrive like mana from heaven after long years of frustration. For "Jonathan" (this is a pseudonym and all names in the remainder of this paper have been changed), a 30-year-old Bayreuth-based academic, the annual application adds deep layers of meaning:

"[I] found the idea of preparing and fighting for something like this for so long actually very nice. Because where else do you do it? And I always think it's nice that way. [...] You have to apply for it every year anew. Um, in the past it was by post, today it's online and if you forget that, you're out of the game. Then you can start all over again. And that's already part of the ritual, the performance. And I was tempted by the fact that

there are limits to get it. That you have to do something for it. And this thinking ahead, I find, a period of 7 years, and back then, it was supposed to take 12 or 13 years, that's kind of absurd too. But as I said, I found it attractive. I don't know why exactly" (Our translation from German language interview).

To Jonathan, the ticketing process is a mixture of a thrill about the gamble to get tickets, despair about the low chances and the hope for a unique experience at the opera. It is also an annual "ritual" that once involved mailing letters and now anxiously waiting for the exact moment the online shop opens while repeatedly pressing the 'return' button on a computer keyboard (a nerve-wracking five minutes we replicated during our fieldwork). When Jonathan succeeds with his application, he is extremely happy. His expenditure of energy will bring truly meaningful rewards. Jonathan is lucky—or perhaps, as we will see, unlucky—because he lives in Bayreuth. Most festival visitors need to travel. This is costly financially, costly to family life and costly in terms of energy.

4. Travel

For those living overseas, the unpredictability of receiving tickets adds serious organizational problems. It is impossible to plan more than a few months ahead. Carol, a 60-year-old Australian member of a Wagner Society, embodies something close to the maximum of sacrifice in terms of distance travelled, financial efforts relative to income, and disruption to her personal life. On being allocated some tickets, she felt she had no choice but to augment the travel plans to Europe that had already soaked up her carefully budgeted vacation savings.

"We had a trip planned to Ireland and Scotland. And that trip was already organized. And then suddenly here I was with an opportunity to go to Bayreuth. Well so that really made a mess of my trip to Ireland and Scotland. So I'm still going. I'm meeting the girls in Dublin on the 30th of August and so the preparation was crazy. Because suddenly I had to factor in an extra two and a half weeks here with my six tickets to Bayreuth Festival and I had to reorganize my life. And I have a mother and a husband and a daughter and so the preparation the logistical the logistics were just insane. But I'm here. I got here."

Carol demonstrates paradigmatically how highly Wagner fans value a visit to the festival and the extraordinary expenses and inconvenient efforts they are willing to take on themselves and impose on others. Most people would consider a holiday in Ireland and Scotland the experience of a lifetime—enough indulgence for one summer. But Carol does not even think about giving away her tickets. As a former teacher, she has a modest income. Yet she spent thousands on new flights and hotel bookings, extending her trip and causing chaos for her family. As a devotee, what else could she do?

After all the planning, there is the actual 'getting there' that involves moving a body through space. This is a major source of suffering for international travelers especially. Since the 19th-century, attendees have complained about the difficulty of getting to Bayreuth and notably the poor train service on peculiar branch lines. Bayreuth still has poor connectivity and visitors flying in from abroad usually come via Munich, then Nuremberg, to the festival. Laura, by coincidence another 60-year-old Australian member of a Wagner Society, offers a complaint that became familiar to us concerning the troublesome, multi-connection route from the airport to Bayreuth (something we have experienced more than once during our fieldwork). True to the pilgrimage motif, she told a story of hurdles and challenges arriving one after another and of a body that had to suffer and take risks.

"But last year the flight was into Munich and then I took the train and I did it all in one hit without stopping until I got to Bayreuth. And that involved a train. I'd been on the plane for goodness knows how many hours 24 h or something and then you get a train a train to the main station in Munich. That worked well. Then I got on a train to Nuremberg and then I had to get off and change platforms to get the train to here. And I had to haul my suitcase. There were no lifts no elevator I had to and I had a bad back. I was not supposed to be doing it. I had to pull my suitcase up the stairs up a really

long flight of stairs in a rush to get to the train. Fortunately, a young woman helped me. The men don't help you and I got to the train and not enough carriages had turned up. Again you've got the dreadful trains with steps to get your suitcase up. You know it's not sliding you've got to go up and I got to there was no room to even get in the carriage. You had to just stand there. So, on top of twenty four hours in the plane and I'd probably by then had at least two and a half hours or something of trains, three hours maybe. I then had that stand for an hour and it was really stuffy and hot. Just in the stairwell not in the carriages for an hour to get here and that was just dreadful . . . "

This is a story of sacrifice in the pilgrimage mode. There is endurance, fatigue, physical effort and even danger to the body.

5. The Performance

What could be more relaxing than sitting in a chair listening to opera? Well quite a few things actually. Once you have your tickets, have made it to Bayreuth and up the hill into the concert hall, the economy of effort and suffering does not stop. There is a tax on the body, on the mind, and the soul. In a letter from 1909, the novelist Thomas Mann wrote to a friend about his summer:

"And then I was in Bayreuth for the Parsifal . . . But quite apart from the physical exhaustion of it all (it was dreadfully hot), I also found it a demanding emotional experience". (Mann 1985, pp. 44–45)

We find a similar response in the composer Tchaikovsky's (1876) report that although he was a "musician by profession," he was "overcome by a sensation of spiritual and physical fatigue close to utter exhaustion" by the end of *Götterdämmerung*, the fourth opera in *Der Ring des Nibelungen*. The Wagnerian George Bernard Shaw was an active young man in the 1880s. Likewise he noted in his dispatches the cognitive labor involved in being a conscientious opera fan and music journalist: "It is desperately hard work this daily scrutiny of the details of an elaborate performance from four to past ten" (Shaw 1889a). The renowned author and critic singled out the particular challenge of the *Meistersinger*, "I had just energy enough to go home to my bed, instead of lying down on the hillside . . . That Third Act, though conducted by Hans Richter, who is no sluggard, lasts two hours, and the strain on the attention, concentrated as it is by the peculiarities of the theater, is enormous" (Shaw 1889b).

What exactly are those 'peculiarities' that troubled Shaw? In contrast to regular concerts, most attendants honor the ritual of the Bayreuth Festival by being dressed in tuxedos or other elegant clothes. These are restrictive and do not permit good ventilation of the body. The central European summer reliably reaches over 30 degrees C. The performances are long, the music tempi generally languid. The doors to the auditorium are closed for acoustic purposes. Yet, even today, there is no air conditioning as the building is heritage listed. The heat is trapped. The famously hard seats help somewhat to keep the audience awake, but in their own way they also make focused attention a challenge. This is all something of an ordeal. Peter, a 67-year-old American living in Germany's Heidelberg, is a first-time visitor. He describes his experience in one of the more affordable parts of the concert hall:

Our seats are way up so it's very hot. And people are we're all packed in like sardines up there. So it's not . . . not comfortable at all but you know you just have to focus.

Peter had done well for a novice—he had translated discomfort into a resource. Yet Helena and Stefanie, two German women aged 70 and 79, underestimated the strain of the long performances, high temperature, and the hard wooden seats. They gave up after the first act due to the unbearable conditions and felt that they could not continue. The trip to Bayreuth was for these ladies a rare opportunity because they were lucky in the online ticket sale. For them, going to Bayreuth was a once-in-a-life-occasion and one of many high-culture events that they had attended. But they were not true adepts. Their motivation was more curiosity than a special fascination with Wagner. Unlike more experienced festival visitors, these mature-age first-time visitors were not well prepared for the uncomfortable seats and slow performances. They paid a price for it. Stefanie:

> *In general, I felt good. But the chairs made it hard for me as said before. And then you lose focus. Then I was looking around so much in the theatre to see who is there. You don't notice anymore what is going on onstage. Yes, I was fidgeting a little. There were two girls whispering. Just like us. She [Helena] wanted to tell me something and I had to stop her. It was like that. One loses concentration. One looks left and right to see what others are doing. That's part of it just to stay flexible. Not sitting still. [. . .] Yes, with other seats we may have fallen asleep but not with these. (Interview in German. Our translation).*

Helena and Stefanie were unable to translate discomfort into an incentive to concentrate. Unlike those who had invested heavily—financially, with travel, with bodily effort, with the ritual preparations of studying the artworks—they were unable to interpret the experience as pleasurable or as offering the possibility of encountering transcendence. Viewed through the lens of the *Elementary Forms*, such draining experiences and privations, forms of disciplinary sacrifice, are perhaps essential as preparations for leaving behind the profane. Hence, Durkheim (1995, p. 314) writes: "no one can engage in a religious ceremony of any importance without first submitting to a sort of initiation". In this context, it is noteworthy that Tchaikovsky (1876), who attended the first iteration of the festival in 1876, considered that his own lack of preparation contributed to a somewhat reserved appreciation of *The Ring*. He noted that other professional musicians were more enthusiastic than he was before admitting, "I am willing to grant that it is my own fault that I have not yet come to appreciate fully this music, and that, once I have got down to studying it diligently, I too may eventually join the wide circle of genuine admirers of Wagner's music".

6. Transcendent Experiences

So attending Bayreuth requires effort and energy. It is draining and uncomfortable. For many, it is expensive. You need to prepare diligently. Expenditures of all kinds abound. But, as both Bataille and Turner argue, at the end of the suffering there can come a connection with sacred spaces and traditions. For example, the reward can be a sense of entering an enchanted land that is separated from profane space. Bayreuth is where the operas that people have been listening to for decades are supposed to be performed—in the building designed by Wagner, built to his specifications, and where the master's eyes personally oversaw the civilizationally significant first productions of the entire *Ring* (in 1876), and *Parsifal* (in 1882). For many fans, it is a place that they have read about for years in books about Wagner or the festival, or seen referred to on program notes at other concert halls. In addition to such mystique and mythology, the very travel involved to this annoyingly remote location builds commitment and contributes to a liminal separation from everyday life. A sense of special, slow time can intersect with that of space. Even our German interviewees told us that going to Bayreuth is a very different experience than visiting a performance in their hometowns such as Berlin or Dresden. Instead of rushing to the theater after the office, most festival visitors dedicate the whole day to opera. In the morning, there is often a presentation at the *Festspielhaus* (the concert hall) that introduces the performance of the afternoon. The day can be spent carefully preparing the mind and body. A study of the libretto is usual. This can be arduous. Alcohol is avoided, food light, an afternoon nap advisable. As in the Aboriginal rituals studied by Durkheim, these smaller preparations, prohibitions and privations are all organized so as to maximize the climactic experience of the artworks over a four-to-eight hour period starting according to Wagner's instruction in the late afternoon.

Several of our interview partners invoked the magic of place, especially during their first visit to Bayreuth. As we just noted, many had spent years reading about Wagner and Bayreuth, or looking at photographs. Encountering them in reality was an intense experience that blurred the lines between myth and reality—not unlike when we encounter a celebrity in the street. When asked about this, Preston, a 60-year-old visual artist from the United States, repeated himself as he struggled for words. He spoke in hushed, breathy tones as he mentally recaptured the feeling of an intense sense of otherworldliness he had experienced when he first visited Wagner's own villa, *Wahnfried*.

> Question: *What about the fact this is Wagner's town? Does that make it more meaningful?*
>
> Preston: *Yeah especially in the beginning. I'm just in awe. Every, everywhere you go everything you see is Wagner. It's a magic, magic Kingdom ... This is Wahnfried! This is the dream. The dream.*

More important even than *Wahnfried* as a sacred space was the *Festspielhaus*. Designed and built by Wagner after considerable financial struggles, it is arguably the largest wooden building in the world. With this all-wood construction, hollow pillars and resonant floors, it has unique acoustics that send vibrations through the body. It seems alive. Some remove their shoes so as to feel this through the soles of their feet. The hall is also famous for the concentration of the audience—known as the "Bayreuth hush". This interactional norm indicating shared devotion also makes the performances special for Preston.

> *Physically the wood and the sound there's no way in the world anywhere could you build this. I keep forgetting I haven't been here for a couple of years. I started coming in 2001. It's really the orchestra and the sound. Also the concentration of the audience it is pitch black and you're packed, packed in this little place and all you see is the performance and you hear the sound and music.*

Another informant, Simon, echoed these feelings about sacred, special places. Simon (German, aged around 50) is today a professional Wagner intellectual and cultural heritage manager. His initial contact with the sacred at Bayreuth came decades ago when he was a student. The visit was literally life changing.

> *"I remember when I first came to Wahnfried at twenty-one years old when I entered Wahnfried I really walked on the top of my toes. It was something very, very holy like coming to the Holy Grail".*

While Bayreuth has sacred places, it is Wagner's music itself that provides a culmination to the pilgrimage quest for the extraordinary. His music has a mesmerizing, deeply emotional effect on many listeners and offers to them a glimpse of transcendence. The impacts can be mysterious and are very hard to explain. Shaw touched upon this theme long ago.

> *"This Parsifal is a wonderful experience: not a doubt of it. The impression it makes is quite independent of liking the music or understanding the poem ... When you leave the theater after your first Parsifal you may not be conscious of having brought away more than a phrase or two of leitmotif ... yet before long the music begins to stir within you and haunt you with growing urgency that in a few days makes another hearing seem a necessity of life".* (Shaw 1889c)

Thirty-six years later, Shaw was awarded the Nobel Prize for Literature. So, it is no surprise that our less verbally talented respondents also struggled to explain what was going on. Steffen (German, aged about 60), an accomplished musician, made use of a metaphor we often encountered (and indeed can be found in Nietzsche's writings)—that of bathing in warm water.

> *"It grabs you by the intellect and grabs you by your feelings and sentiment Both, right? You're swimming in beautiful, warm water. "*
>
> Sandra [Steffen's wife] added that *"My husband can't talk for a long time after such a ... It's, uh ... (Interview in German. Our translation).*

This shift towards introversion and silence that Sandra noticed happened to us too. In our field observations, we often noticed that other audience members were unusually quiet as they filed out during intermissions or at the end of the long evening. It seemed as if the music had cast a spell and it would take a while to return to a profane world where words were adequate to experience and interactions anything other than a crude intrusion. Like Thomas Mann and Steffen, we and they were emotionally full and emotionally exhausted at the same time. Chat would typically resume as patrons unlocked their cars or passed by the railway station, after about a kilometer on the long, late-night downhill walk back into town.

True to Durkheim's vision of the sacred as a commanding force, Steffen went on to describe the moment of ecstatic submission to the power of the music with reference to the climax in the Third Act of the *Meistersinger*, where various contending keys that have been at play for the prior four or five hours resolve to C major. The impact on the audience, the vast majority of whom would not be intellectually aware of the compositional technicalities, was such that *"it knocks them out. It knocks them out. It brings tears to their eyes or something. They can't breathe."*

This sense of meaningful contact with intense emotions and a higher power was also captured by Preston, who contrasted the unique impacts of Wagner to those of other composers. Usually articulate, Preston really struggled to communicate his feelings during our interview.

> *"I play a lot of Beethoven, Bach, Mozart everything. But with Wagner. He has a knack of taking control of you. Me anyway. His music almost grabbed hold of you and just he could . . . manipulate you. And I just, you just totally surrender to Wagner. Other people like Beethoven you play through and there's no emotion, I don't feel any anything inside. With Wagner it's different. And as older when I get older it's even worse you know because . . . with Parsifal the music. At one point when the Gurnemanz is singing especially if it is a good bass singing . . . and with Wagner's music. I completely lose control . . . to tears."*

Simon's experience was even more intense. We last encountered him entering *Wahnfried* on tip-toe. Later that same day, he had slipped into the *Festspielhaus* and listened to a dress rehearsal of *Die Walküre (Valkyrie)*. His account of what followed was full of the repetitions and expressive struggles we often found in our transcripts when respondents attempted to communicate a sense of the transcendent.

> *"It was like an infection like a medical infection. I remember after, after Valkyrie I had a spare day, a free day after Valkyrie. I attended the general uh the dress rehearsals and after Valkyrie I had a free day and I took a walk up on the on the Bürgerreuth, this hill behind the Festspielhaus. I climbed up to the Siegesturm, this tower of victory. Yeah I was standing there quite alone for myself and I was in a very, very special mood. It was like everything was floating through me and I was fascinated and really was crying. It was a deep, deep emotional effect. Really it was erotical in a way and very, very special. And well it was like . . . was ist ein Erweckungserlebnis? An awaking experience. Suddenly something became light in my head. And if it was like like this uh this flash of Erkenntnis, of knowledge, of knowledge. Everything seemed to be clear and I had the impression that I had found something very important for me in my life. Okay. That was the infection and I couldn't stand it anymore."*

Simon had indeed encountered the sacred as he looked out over the landscape, the town, fields and rolling wooded hills. There is perfect consistency with Durkheim's observation in the *Elementary Forms* about the intense experience of a higher power after ritual:

> "Its immensity overwhelms him. That sensation of an infinite space surrounding him, of an infinite time preceding and to follow the present moment, of forces infinitely superior to those at his disposal, cannot fail to arouse the idea inside him that there is an infinite power outside him to which he is subject. This idea then enters into our conception of the divine as an essential element." (Durkheim 1995, p. 80)

It is an outcome of which Wagner, no doubt, would have approved.

7. Some Closing Thoughts

We have written about participation in an event that is an impractical, expressive expenditure of surplus, not a rational, accumulative, profitable enterprise. Tchaikovsky (1876) captured this essence when he wrote: "In the sense of contributing to the material prosperity of mankind, the Bayreuth Festival, of course, is of no consequence whatsoever, but in the sense of a quest for the realization of artistic ideals it surely is fated, in some way or other, to acquire a tremendous historical significance". The pilgrimage to Bayreuth is a particular kind of sacrificial activity that attaches to this broader aesthetic project. Suffering, travel to a sacred space, and mental and physical discipline of listening lead in some cases to a transcendent encounter with deeply meaningful complex cultural forms. It is a serious and contemplative mode of expenditure that provides only spiritual rewards. And that is the entire point.

The sacrifice of time, commitment and money is the required expenditure of surplus that makes the trip to Bayreuth a maximal spiritual success. We saw how Wagner fans suffer before and during the events. In some cases, these visitors are completely overwhelmed by attending the operas. The aesthetic experience of the musical theatre, the magic of the place, and restricted access to one of the sought-after performances combine alchemically. The several-hours-long operas, the heat in the Franconian high summer and concentration on the demanding art works distinguish a trip to Bayreuth from fun trips to pop concerts or city breaks. But even as people complain about them, these hurdles are in fact a necessary step towards a more transcendent experience. With the right preparation and mind set, something truly life changing can happen. Within the framework of Bataille's understanding of sacrifice, the impacts of these non-utilitarian costs and privations make perfect sense. As an act of sacrifice built on the hope of grasping transcendence, each of the seemingly irrational acts of spending and restraint brings participants a little closer to their goal. Bataille perfectly sums all this up, not with reference to Europe or modernity but in his understanding of sacrifice among the Aztecs:

> "The only valid excess was one that went beyond the bounds, and one whose consumption appeared worthy of the gods. This was the price men paid to escape their downfall and remove the weight introduced in them by the avarice and cold calculation of the real order." (Bataille 1988, p. 60)

The crucial point about human sacrifice for Bataille was that it tore down the established profanity of everyday life and put the individual in immediate contact with the sacred and divine. In its expressive details, an opera in a Bavarian province is not the same as the religious ceremony of an extinct Mexican culture. However, they are analytically akin. Excessive efforts, incommensurability beyond the exchange of equivalents, and the struggle to escape the profane world are the carriers of symbolic and emotional power in both the Aztec event and the German opera festival.

Bataille's brutal case study can be considered a kind of Foucaultian real-world exemplar of a system of ideas and rituals. It highlights by looking to the extreme In Bayreuth, the cultural patterns are a little more muted and veiled. So of course, there will be a range of experiences and levels of commitment. As we saw with Helena and Stefanie, not everyone going to the opera is an adept or enthusiastic Wagnerian or can be easily interpreted as engaging in pilgrimage activity. For some locals we interviewed, a visit to the *Festspielhaus* is just another entertaining night out—a lesser form of sacrifice to be sure but for all that an irrational expenditure of efforts just the same. Nor is all tourism to Bayreuth driven by Wagner. Many come to enjoy the beer, the landscape, the flower gardens, and the castles and stately homes that litter the region. And in the everyday life of the town's residents, we might find shopping, drinking, eating, hobby activity, physical training, reading, focused attention, and sex. But are these not also gratuitous expenditures? Understood through the lens used in this paper, we might reconstruct a more comprehensive, nested and layered ecology of the expenditures and sacrifices within the city of Bayreuth, of which the festival and its 'pilgrims' are but a seasonal part. Foucault spoke of a vision of the carceral city, of a network of institutions and activities of control spread out over urban space. With a very different *dispositif* of modernity, a Durkheimian cultural sociology could map out the town's landscape of individual and institutional discipline, effort, sacrifice, emotional and expressive release. To do so would be a magnificent accomplishment, and clearly one that is beyond the scope of this initial inquiry. It is, however, an agenda called forth by the maximal understanding of sacrifice as the expenditure of surplus. Our sketch of the Bayreuth pilgrimage is just an illustration of how one might go about such a task.

Author Contributions: The authors contributed equally to the research and writing of this paper. Both authors have read and agreed to the published version of the manuscript.

Funding: This research was supported by a grant from the MacMillan Center of Yale University and by a University of Bayreuth Centre of International Excellence "Alexander von Humboldt" Senior Fellowship.

Institutional Review Board Statement: This research was conducted according to the Institutional Review Board and informed consent protocols of Yale University and Bayreuth University.

Informed Consent Statement: Informed consent was obtained from all subjects involved in the study.

Data Availability Statement: The authors have all data available. However, the data are subject to restricted access due to confidentiality.

Conflicts of Interest: The authors declare no conflict of interest.

References

Adorno, Theodor. 2005. *In Search of Wagner*. London: Verso.

Alexander, Jeffrey, and Philip Smith. 2010. The Strong Program. In *Handbook of Cultural Sociology*, 2nd ed. Edited by John R. Hall, Laura Grindstaff and MingCheng Lo. Abingdon: Routledge, pp. 13–24.

Bataille, Georges. 1985. *Visions of Excess*. Minneapolis: University of Minnesota Press.

Bataille, Georges. 1988. *The Accursed Share*. New York: Zone Books.

Collins, Randall. 2004. *Interaction Ritual Chains*. Princeton: Princeton University Press.

Durkheim, Emile. 1995. *The Elementary Forms of Religious Life*. Translated by K. Fields. New York: Basic Books.

Evans-Pritchard, E. E. 1964. Foreword. In *Sacrifice: Its Nature and Function*. Edited by Henri Hubert and Marcel Mauss. Translated by W. D. Halls. London: Cohen and West, pp. vii–viii.

Gebhardt, Winfried, and Arnold Zingerle. 1998. *Pilgerfahrt ins Ich: Die Bayreuther Richard Wagner-Festspiele und ihr Publikum*. Eine kultursoziologische Studie. Konstanz: Universitätsverlag Konstanz.

Heller, Erich. 1985. Introduction. In *Thomas Mann Pro- and Contra Wagner Op-Cit*. Chicago: University of Chicago Press, pp. 11–22.

Hertz, Robert. 1913. Saint Besse: Etude d'un culte alpestre. *Revue de l'Histoire des Religions* LXVIII: 115–80.

Hertz, Robert. 1960. *Death and the Right Hand*. London: Cohen and West.

Hubert, Henri, and Marcel Mauss. 1964. *Sacrifice: Its Nature and Function*. Translated by W. D. Halls. London: Cohen and West.

Macclancy, Jeremy. 1994. The Construction of Anthropological Genealogies: Robert Hertz, Victor Turner and the Study of Pilgrimage. *Journal of the Anthropological Society of Oxford* 25: 31–40.

Malinowski, Bronislaw. 1932. *Argonauts of the Western Pacific*. London: Routledge.

Mann, Thomas. 1985. *Pro- and Contra-Wagner*. Chicago: University of Chicago Press.

Mauss, Marcel. 1954. *The Gift*. Translated by I. Cunnison. London: Cohen and West.

Miller, Daniel. 1998. *A Theory of Shopping*. Cambridge: Polity.

Riley, Alexander T. 2013. *Godless Intellectuals*. New York: Berghan Books.

Robertson Smith, William. 2002. *Religion of the Semites*. New Brunswick: Transaction Books.

Shaw, George Bernard. 1889a. Impressions de Voyage. *The Star*, August 2.

Shaw, George Bernard. 1889b. Bayreuth and Back. *The Hawk*, August 13.

Shaw, George Bernard. 1889c. The Second Parsifal. *The Star*, August 7.

Smith, Philip. 2020. *Durkheim and After: The Durkheimian Tradition 1893–2020*. Cambridge: Polity Press.

Spotts, Frederic. 1994. *Bayreuth: A History of the Wagner Festival*. New Haven: Yale University Press.

Tchaikovsky, Pyotr Illyich. 1876. The Bayreuth Music Festival. Available online: http://en.tchaikovsky-research.net/pages/The_Bayreuth_Music_Festival (accessed on 18 December 2020).

Turner, Victor. 1969. *The Ritual Process*. Chicago: Aldine.

Turner, Victor. 1973. The Center out There: The Pilgrim's Goal. *History of Religion* 12: 191–230. [CrossRef]

Wagner, Richard. 1994. *Religion and Art*. Omaha: University of Nebraska Press.

West, Brad. 2008. Enchanting Pasts. *Sociological Theory* 26: 258–70. [CrossRef]
Zinnecker, Florian. 2008. Süddeutsche Zeitung Bayreuther Kartentricks. Available online: https://sz-magazin.sueddeutsche.de/musik/bayreuther-kartentricks-75596 (accessed on 8 August 2020).

Article

The Dark God: The Sacrifice of Sacrifice

Joseba Zulaika

Center for Basque Studies, University of Nevada, Reno, NV 89557, USA; zulaika@unr.edu

Abstract: The Frazerian question of murder turned into ritual sacrifice is foundational to cultural anthropology. Frazer described the antinomian figure of a king, who was, at once, a priest and a murderer. Generations of anthropologists have studied sacrifice in ethnographic contexts and theorized about its religious significance. But sacrifice itself may turn into a problem, and René Girard wrote about "the sacrificial crisis", when the real issue is the failure of a sacrifice that goes wrong. The present paper addresses such a "sacrificial crisis" in the experience of my own Basque generation. I will argue that the crisis regarding sacrifice is pivotal. But my arguments will take advantage of the background of a more recent ethnography I wrote on the political and cultural transformations of this generation. This requires that I expand the notion of "sacrifice" from my initial approach of ethnographic parallels towards a more subjective and psychoanalytical perspective. As described in my first ethnography, the motivation behind the violence was originally and fundamentally *sacrificial*; when it finally stopped in 2011, many of those invested in the violence, actors as well as supporters, felt destitute and had to remodel their political identity. The argument of this paper is that the dismantling of sacrifice as its nuclear premise—the sacrifice of sacrifice—was a major obstacle stopping the violence from coming to an end.

Keywords: sacrifice; martyrdom; ETA; Yoyes; ethnography; psychoanalysis

Citation: Zulaika, Joseba. 2021. The Dark God: The Sacrifice of Sacrifice. *Religions* 12: 67. https://doi.org/10.3390/rel12020067

Academic Editor: Javier Gil-Gimeno

Received: 21 December 2020

Accepted: 12 January 2021

Published: 20 January 2021

Publisher's Note: MDPI stays neutral with regard to jurisdictional claims in published maps and institutional affiliations.

1. Sacrifice as Duty and Crisis

Sacrifice is a central topic in modern anthropology. Frazer addressed it while reporting on the institution of divine kingship found in many ethnographic societies; typically, when a king became old and feeble, the future monarch would challenge him to a duel, kill him, and take over the priestly and political powers of the dead king. Thus, he was, at once, "a priest and a murderer" (Frazer 1963, p. 1). Evans-Pritchard expanded this ritual complex to the study of Shilluk regicide (Evans-Pritchard 1963). Recently, Sahlins and Graeber (2017) have revisited and updated the theoretical foundations of kingship and sovereignty. René Girard argued that, given the absence of a judicial system in primitive societies, sacrifice was a key form to restrain vengeance—"an instrument of prevention in the struggle against violence" (Girard 1977, p. 17). Maurice Bloch examined how ritual achieves transcendence by the sacrifice of the participants, thus affirming through symbolic violence the timeless truth that binds a community to a belief or a cause (Bloch 1992, 2013). Based on the principle that violence and the sacred are inseparable, sacrificial rites assume essential functions in restoring social control. I applied this ritual model, in which sacrifice and murder substitute reciprocally, in my own study of the Basque political violence of the 1970s. Following Roy Rappaport's statement that "Morality, like social contract, is implicit in ritual's very structure" (Rappaport 1979, p. 198), the aim of my ethnography was to show the cultural, performative and religious dimensions of the violence. Still, despite all the models I borrowed from ethnography and the literature, I concluded that "The thing itself, the sacramental literalness of the sacrificial act, cries out against any final interpretation" (Zulaika 1988, p. 342).

Two decades after the end of the Spanish civil war, dictator Franco was still in power when, in the 1950s, a small group of young, politically minded students began meeting in Bilbao to study Basque history and language. One of them was Julen Madariaga, a member of a prominent Bilbao family of lawyers who had returned after a decade of exile in Chile; he was three years old when was exiled as Franco's army was closing in during the Spring of 1937, days before Guernica was burnt to the ground by Hitler's planes. When, in the summer of 1959, they founded the underground ETA (*Euskadi Ta Askatasuna*—"Euskadi and Freedom"), Guernica's sacrifice was the axiom behind the group's commitment. They carried in their pockets a small volume known as *The White Book* (due to the color of its cover), which summarized their cause. Its insistence on the primacy of "conscience" and "responsibility", quoting Catholic moralists such as Maritain, sets it closer to the *Spiritual Exercises* of the Basque founder of the Jesuits, Saint Ignatius of Loyola, than to anything of their own day. The ideological pillars of the new patriotism were an irrevocable Ignatian decision to surrender one's will for the cause—Sartrean absolute freedom—and the founder of Basque nationalism's Sabino Arana's oath to offer one's life for the fatherland. ETA's initial mission was to create a new subject capable of total sacrifice in the fight against Franco's regime.

There were two prominent figures in literature and art in the Bilbao of the 1950s—Ernest Hemingway, a frequent visitor to the city's summer bullfights, and the local sculptor Jorge Oteiza. "Art is sacrament" is the logo that condenses Oteiza's thinking in a book he completed in Bilbao in 1952. "Writing is tauromachy" is the equivalent summary of Hemingway's work, the American writer who commanded Bilbao's largest international audience. Yet, at the turn of the 1950s, both men were experiencing an existential crisis. Their culture of sacrifice and sacrament had turned down a blind alley. Oteiza quit sculpting in 1959; Hemingway committed suicide in 1960. They were both representatives of what philosopher Maria Zambrano named "the generation of the bull", people who gave it all in their fight against fascism, "because of their sense of sacrifice" (Zambrano 1995, p. 44). Picasso's sacrificial tauromachy for "Guernica" is the emblem of this generation.

There was a category of young people who were particularly attuned to ETA's sacrificial politics—seminarians and religious people. Hundreds of them left the seminaries and joined ETA's ranks in the 1960s. During Franco's era, education was, for the most part, in the hands of religious orders, which meant that, for most lower-class people, the only possibility of a secondary education was internment in a seminary or convent. In the process of schooling, religious institutions would fish for "vocations" for priesthood. A critical part of the indoctrination was the duty of sacrifice for the sake of one's own salvation and the world's redemption. As if the daily sacrifices of religious discipline, endless prayer, and even self-flagellation were not enough, a favorite fantasy of these orders was martyrdom in some faraway missionary post, which was to be embraced as an ardent desire and a secret enjoyment.

But such religious idealism could not endure confrontation with the reality of contemporary life. Authors such as Nietzsche played a key role in awakening this generation to the profound nihilism behind a passion for sacrifice that was "*a will to nothingness*, an aversion to life, a rebellion against the most fundamental presuppositions of life" (Nietzsche 1967, p. 163). Seminaries and convents were mostly empty by the late 1960s. With the loss of the religious world, its entire system of beliefs and values became meaningless, which meant that there was no longer any reason for sacrifice. Far from feelings of exhilaration for the new freedom, the common experience was rather a sense of vacuity and meaninglessness. Loss of faith meant a denial of the big Other of religion—a lack of belief, a lack of commitment, and a lack of sacrifice. It was disbelief *after* belief, de-conversion *after* conversion. In Hegelian terms, it was the negation of negation, or the redoubling of reflection by which the subject posits their own presuppositions. It was the sacrifice of sacrifice, experienced in a state of subjective destitution (Zulaika 2014).

But this overturning of the duty to sacrifice was easier said than done. The religious desire of self-immolation may have transformed into some form of delirium, but was politics not the *real* domain where sacrifice made sense in the fight against military fascism? After the general emptying of seminaries and convents, those hungry for sacrifice had a legitimate substitute in surrendering to a political commitment that demanded the perilous rigors of underground activism, which included armed action. The consequences of such a fateful decision were almost certainly torture, death or exile.

It was one thing to give up religion as a fundamental fantasy, but quite another to give up what we might call, in psychoanalytic terms, its *infinitization* of desire, for "desire is nothing but that which introduces into the subject's universe an incommensurable or infinite measure" (Zupancic 2003, p. 251). This is a desire that resides in "the body as distinct from the organism inasmuch as it is not a biological real but rather a form" (Miller 2009, p. 40). No longer able to believe in religious transcendence, many of my generation decided to surrender their lives to the political cause—"giving your life" for political freedom was a way to repeat that fullness of sacrifice in the infinitization of desire. For the many former seminarians and priests who entered ETA, the opportunities of new political martyrdom it offered were a thousand times more preferable than the destitute emptiness of a world without a Cause.

ETA filled the passion for sacrifice to the brim. Still, the nationalist desire regarding the Cause for which the youths were ready to die was not the same as that of the older generation; the structure of transference for the Basque nationalists who fought the Spanish civil war of 1936–1939 had been Arana's formula for sacrifice: "Me for Euskadi and Euskadi for God". One of the victims in Guernica was "Lauaxeta", a well-known poet who was arrested and executed; before facing the firing squad at dawn, he wrote a farewell poem to his country, which concluded: "Let the spirit go to luminous heaven/Let the body be thrown to the dark earth". In the ETA of the 1960s, such religious mediation had no place in the ideology of the militants who, by the end of the decade, were, for the most part, avowed Marxists and atheists. For the older generation, the nationalist duty to fight for their country had the Homeric inevitability of defending one's community militarily from the antidemocratic forces of European fascism; for the new ETA generations of youths, it was a much more individualized call; their notion of "freedom" was far more personal and political, and mediated as much by the writings of Marx, Nietzsche, Sartre and Dostoyevsky than those of Arana or the Bible.

2. Sacrificing Your Love to "Freedom or Death"

Together with Madariaga, the other most influential founder of ETA was José Luis Álvarez Enparantza, "Txillardegi". During the formative years of ETA. Txillargedi published a novel in 1957, constituting a breakthrough in Basque literature because of its existentialist themes staged in urban settings. Txillardegi, ETA's main ideologue at the time, admitted that this novel was a testament to the subjective and intellectual issues of the period. One theme shown in the novel is the direct influence of political militancy on the sex lives of the protagonists. In the convents and seminaries, there was little doubt regarding the issue: strict chastity was the rule. There was no such rule in the area of politics, but many of the male activists would behave according to the premise that their patriotic duty was above any love affair and that "the sacrifice of the woman" was to be expected.

Txillardegi's novel narrates the failed love between the protagonists Leturia and Miren, which ends in her sickness and death, followed by his suicide. After marrying Miren, and then abandoning her to go to Paris and experiment with a life of his own, Leturia comes to the realization that "my heart needed something Absolute", and falling in love with a woman was only a symptom of that need. Leturia debates the conflicting demands that derive from his reason (which makes the subject the center of his world) and his sentiment (which demands the surrender of one's life for others). Leturia finds a resolution in *The Tragic Sense of Life* (Unamuno 1972), the Bilbaoan philosopher who "denies to thought the capacity to find truth", in Txillardegi's words, "and he takes the road of sentiment alone" (Alvarez Enparantza 1985, p. 73). Unable to choose, Leturia thinks that the best thing he can do is surrender to destiny; he describes his relationship to fate with the analogy of the dog in relation to his master: "I have to ask not 'Who is my servant?' but 'Whose servant am I?' This is the salvation". Salvation is serving the big Master. In the end, Leturia's love for Miren is a thinly veiled metaphor for his love of the Motherland: "Miren needs me; my motherland needs me; my people need me. I belong to them, and without them I am nothing". He recognizes, "I am guilty, yes", for having abandoned Miren/the Motherland, and he promises that, if she survives, "I will redeem my sins with love. With love ... with love ... Who used to speak always about love? I am afraid to admit it: Christ!" (Alvarez Enparantza 1977, pp. 137, 139, 142). As in the plot of a Greek tragedy, one is *guilty* whether one sides with the law or fights against it. Leturia is barely a Christian, and Txillardegi's protagonist in his next novel (published in 1960) is no longer one. The previous generation of Basque nationalists who fought the war were guided by the slogan "God and Ancient Laws". For the ETA generation, God was no longer the big Other. Even though Txilardegi's Leturia was inspired by atheists such as Unamuno and Sartre, he repeats Christ's formula as his mantra: "And what is love? To lay down your life" (Alvarez Enparantza 1977, p. 136). In short, ETA members were Sartrean existentialists, Dostoyevskian nihilists and Nietzschean atheists, but their supreme model of *passage à l'acte*, the only one their audience could totally understand, was none other than the Crucified's sacrificial "gift of Death" (Derrida 1995).

Txabi Etxebarrieta is arguably the most consequential figure in the history of ETA. He joined the armed group in 1963 and came to define its basic ideology as "a Basque socialist movement of national liberation". He was the first ETA member who killed a Spanish policeman at close range and who was subsequently killed by the police himself. But what is most remarkable about Etxebarrieta the writer is his poetry. Before he died in 1968 at the age of twenty-three, he had written five short books of poems—many of them love poems. Right before joining ETA, he had an intense love affair with a woman named Isabel, which was reflected in his poems and letters. Etxebarrieta's older brother, Jose Antonio, was, at the time, a top ideologue in the organization and a mentor to Txabi; he contracted a grave illness and Txabi, besides nursing him, replaced him in ETA's next general assembly, reading a report written by Jose Antonio. Soon, Txabi communicated his decision to join ETA to Isabel, and his letters began to reflect the difficulties in their relationship. At one point, his letters become a repetition of strings of "I love you" followed by "forgive me". One of his poems at the time is marked with the repeated uncertainty of "Perhaps ... ": "Perhaps ... /I am cruel—for committing suicide ... /and for not leaving my blood to others, still unborn". Even for an existentialist nihilist like himself, his passion for suicide was perhaps too cruel. Why did he have to sacrifice himself by sacrificing Isabel? What did she have to forgive him for? Did his motherland and his brother deserve that kind of love?

The Etxebarrietas lived on the fifth floor of an apartment house at the plaza Unamuno in Bilbao's *Casco Viejo*. The plaza displayed the bronze head of Unamuno at the top of a column. From the balcony of the apartment, Txabi would stare at Unamuno's bust and the roofs of Bilbao's old quarters. Unamuno's thinking, best known for his "tragic sense of life", was a major influence on Etxebarrieta. Still, what is most remarkable in Etxebarrieta's early writing is his critical assessment of Unamuno's work. In an essay entitled "Unamuno, Tomorrow (A Feeling Not Felt)", written when he was nineteen, Etxebarrieta distances himself from the Unamunian tragic sense of life in favor of Sartrean existentialism. He pointedly criticizes Unamuno for being one of those who "displace their existential center" toward the future "and are not 'in themselves,' but come to live at the service of the hoped-for transcendence. Only the one who does not expect to be can be 'in himself' comfortably and fully, without any violence". And he continues accusing Unamuno of living in a "dative" mode, i.e., toward an indirect third person or object or temporality—"the today in and for tomorrow, the now in and for later". Etxebarrieta claims that "one should live in a strictly human dimension", an attitude that "dispenses with totalitarian and global solutions" and dismisses Unamuno's tragic sense as "a romantic idea in the irrational sense of the term", because "there is a short step from irrationalism to fascism" (Lorenzo Espinosa 1996, pp. 162–63, 165). Etxebarrieta was essentially saying that there is no big transcendent Other for whom one should live in a "dative", third-person form of indirect subjectivity.

But, again, this was easier said than done. In the end, Etxebarrieta could not live up to his insight; he would be devoured by his own idealism and his passion for sacrifice—sacrifice being the act that "proves" the existence of the big Other. One of his poems, "Motherland", begins with an epigraph from Blas de Otero: "Wretched whoever has a motherland and that motherland obsesses him as much as she obsesses me" (Lorenzo Espinosa 1996, p. 76). Patriotism was both Etxebarrieta's fate and the curse that would not allow him to enjoy a life with Isabel. He writes a poem dedicated to "Your Body": "I'd like to be buried in you./No longer to be" (Lorenzo Espinosa 1996, p. 87). In the last poem of his final book, written just months before his death, an Etxebarrieta now fully surrendered to political action expresses his wish that he could trade it all to simply be Isabel's lover: "With fury I would trade our lives/for the enormous marching of bodies, where loving you would cover me/like the sea covers itself, entirely" (Lorenzo Espinosa 1996, p. 120).

It was not for nothing that Etxebarrieta had long been obsessed with death, which he had repeatedly prophesized for himself. There are repeated mentions of death as a self-fulfilling prophecy in his poetry. In 1965, already in ETA, he wrote three short stories in which the main character has a premonition of death and leaves a farewell to his mother in his notebook (Lorenzo Espinosa 1994, pp. 231–32). In his last political text, written for the occasion of May Day, 1968, and in one of those clandestine cyclostyled pamphlets, Etxebarrieta wrote: "Any day now we will have a dead body on the table"

His day of sacrifice came on June 7 of that year, when he was stopped for a traffic violation. As a policeman, José Pardines, began checking his license plate, Etxebarrieta shot and killed him from behind. He and his ETA companion hid for a few hours in the apartment of an acquaintance. But an agitated Etxebarrieta recklessly decided to leave the hideout, despite the fact that all roads were under police surveillance. He was stopped by police and killed on the spot; his companion escaped but was arrested the next day. His closest ETA friends believe to this day that he let himself be killed (Uriarte 2005, p. 90). One of his ETA comrades wrote, after his death, that he would say frequently, "the country needs me and I will offer myself for her" (Lorenzo Espinosa 1994, p. 271). Another comrade wrote, "at bottom I always thought that he was obsessed with his own martyrdom and perhaps he shot Pardines just so that they could kill him" (Onaindia 2001, p. 322).

Thus Etxebarrieta ended up repeating, in real life, the very subjective structure sketched by Txillardegi in his Leturia character—encountering and falling in love with a woman; the impossibility of maintaining intimate relationships because of the call of the motherland's Absolute; inner struggle between the more rational and the more emotional aspects of their personality; and a decision to take action and the redemption of unconscious guilt for sacrificing the woman through the patriotic love of self-immolation. Both Leturia's narrative and Etxebarrieta's life illustrate the struggle between the body "as the site of *death*" versus the body as "the site of *sex*" (Copjec 2002, p. 28)—a struggle that, as we will see below, would find a historic resolution in Yoyes. Kierkegaard, who wrote the story of Isaac losing his faith in the God of his idolatrous father, was behind much of Unamuno's existentialist thinking, and both philosophers were cornerstones to Txillardegi and Etxebarrieta. What was said of the Danish philosopher—that "Abraham was not only Kierkegaard's father, who offered his son as a sacrifice, but Abraham was also Kierkegaard himself, who sacrificed Regine" (Garff 2000, p. 256)—could be applied to Txillardegi's Leturia character and to Etxebarrieta, both of whom sacrificed what they most loved, Miren and Isabel, for the sake of the country.

ETA's defining alternative, emblazoned as a logo in every pamphlet and publication, was *Askatasuna ala Hil* (Freedom or Death) or *Iraultza ala Hil* (Revolution or Death). Only death could "prove" one was truly fighting for freedom and revolution. But even if ETA's revolutionary discourse spoke of readiness to sacrifice one's life, on that day in 7 June 1968, the haunting issue for Etxebarrieta was how he and the public should view his killing—was it a revolutionary act or a vulgar murder? There was one thing that could demonstrate he was not a common killer—the *sacrifice* of his own life as a proof that he acted for his commitment to the big Other of the patriotic cause. Whether or not his obviously reckless behavior could be construed as "suicide", the political and ethical justification of this inaugural killing and death demanded that it be deemed a revolutionary self-immolation. Etxebarrieta, who saw, as no one else did, the romantic trap of Unamuno's "tragic sense of life", became the paradigmatic figure whose killing and death ironically marked the birth of ETA's tragic period.

The new ETA, erected in the memory of Etxebarrieta's murder/martyrdom, required vengeance from his orphaned comrades. They assassinated a police officer known to be a torturer, Meliton Manzanas. Killing was now a revolutionary demand. ETA was no longer able to distinguish (with Benjamin) between "mythic violence" and "divine violence". Žižek explains: "It is mythical violence that demands sacrifice, and holds power over bare life, whereas divine violence is non-sacrificial and expiatory ... [it] serves no means, not even that of punishing the culprits and thus re-establishing the equilibrium of justice" (Žižek 2008, pp. 199–200). ETA and its followers thought that the premeditated killing of a policeman, to be followed by hundreds of similar killings in the future, was nothing but the logical conclusion of the revolutionary embrace of violence. After Etxebarrieta, it was too late to stop the cycle of human sacrifice. It would take nearly half a century to confront the defeat of such mythic violence.

3. The House of the Father: Patricide, Masochism and the Superego

Etxebarrieta's father died when he was thirteen. The Etxebarrieta family house was in the coastal Bizkaian town of Ispaster. After the war, his family house was turned into a Spanish police station (Lorenzo Espinosa 1996, p. 109). One can hardly think of a greater political offence for a nationalist family. The slogan that summed up the ETA generation's political ethos was the best-known line of the greatest post-war poet, Gabriel Aresti: "I will defend the house of my father". This was the father who had lost the war against fascism and, paradoxically, the sons of the new generation had to kill him first, before defending his house. The same Basque nationalist leaders who had been internationally lauded for their fight against fascism during the Spanish war had become, for ETA's generation, politically irrelevant in their Parisian exile—they were the impotent father who had sold his soul to bourgeois placidity. The Etxebarrieta brothers considered the gradualist approach and struggle of the older generation of nationalists a "crisis of adolescence" (Etxebarrieta 1999, p. 136).

In Txabi Etxebarrieta's essays, one particular writer is quoted and invoked prominently—Feodor Dostoyevsky, the author of the struggle between Christianity and faithlessness, crime and punishment, freedom and guilt. Etxebarrieta mentions each of the brothers Karamazov in his writings. What was fiction in Dostoyevsky's novels a century later became the vividly experienced reality for ETA's generation. But, in the case of *The Brothers Karamazov*, a novel that deeply impacted Etxebarrieta and other prominent Basque writers, one should pay attention to its central plot: it was the story of a patricide. Dostoyevsky put his finger on some of the core issues of Etxebarrieta's generation: the religious killing of their Christian Father, as well as the political killing of their vanquished fathers—patricides that filled them with an unconscious guilt that could only be redeemed by the masochistic passion of religious and political self-sacrifice.

Another author for whom patricide was at the center of his work was Freud. The myth put forward by Freud in *Totem and Taboo* describes a despot father who appropriates all the women of the tribe for himself and who is murdered and his body eaten by his sons. The sons feel remorse for the murder and establish a new order based on the two taboos of exogamy (against the incestuous possession of women) and totemism (against killing the totem animal that, while representing the father, establishes affiliation and can only be sacrificed to divinity). The paradigm of the Freudian sacrifice is "The totem meal", which is, for Freud, "the beginning of many things—of social organization, of moral restrictions and of religion" (Freud 1950, p. 142). Freud writes about sacrifice in relation to "civilization and its discontents"—the fact that civilization is based on controlling instinctive drives, a renunciation sanctioned by religion as a sacrifice to divinity, and which is also, for Freud, the origin of neurosis.

Freud links the primal myth with the structure of Christianity, where "There can be no doubt that the original sin was one against God the Father" (Freud 1950, p. 154)—a myth that, as Lacan adds, "is the myth of a time for which God is dead", a God who "has never been the father except in the mythology of the son" (Lacan 1992, pp. 177, 307) who is commanded to love him. Freud uses this myth to state that, psychoanalytically, in the structure of the unconscious, the father function has to do with castration (in the son's rivalry for the mother's love) and with the origin of the superego. When asked his name, this God/Father responds in the Exodus, "I am who I am"—His name is The Name, and it is in the Name of the Father, turned superegoic after his death and by assuming his voice, that the subject speaks and acts (Lacan 2013, pp. 78–81). The first thing that the believer must do in the Name of the Father is the sacrifice of Isaac, so that the bond between father and son becomes binding.

The birth of Freudian psychoanalysis was related to the inevitable decline of paternal authority in modern societies (Roudinesco 2016, pp. 213–17, 284, 369). This is linked, in our case, to the theme of so-called "Basque Matriarchalism" (Ortiz-Osés 1980). In a nutshell, this was the proposition that women dominate the Basque household and, more relevantly, the Basque unconscious. Basque feminist anthropologists were strongly opposed to this thesis (Del Valle et al. 1985; Bullen 2003). The mediation of the maternal figure of God was prominent in the Marian version of Catholicism typical among Basques. Th structure of desire in such a religious complex is one of maternal sublimation and filial sacrifice. The cult of the Virgin Mother has elements of medieval courtly love in which the Lady operates as a mirror upon which the vassal projects his idealized wishes. What matters is the inaccessibility of the object by which the vassal turns what is an impossibility into a prohibition, the object of desire being the same condition that forbids its obtainment. Sacrifice goes hand in hand with the secret enjoyment of the love object. The male masochistic dream of sacrifice to an idealized woman is summed up by Deleuze in three words: "cold-maternal-severe", where cruelty is intimately related to the Ideal. The guilty masochist asks to be beaten, but for what crime? Deleuze suggests that "the formula of masochism is the humiliated father" (Deleuze 1989, pp. 51, 60–61). The masochist experiences the symbolic order (of religion, patriotism, the family) as a maternal order: it is the Mother who requires the Son's sacrifice. In this cultural configuration, which is constitutive of the ETA generation's subjectivity, masculinity is embodied in the role of the son, whereas femininity is projected onto the role of the mother. Sociologists and historians of ETA have underlined the prominence of mothers in the lives of their militant sons.

The link between the Freudian superego and the demand for sacrifice requires special consideration. The psychoanalytic literature has translated Kant's categorical imperative into the agency of the superego, which is never satisfied and which demands more sacrifice the more we sacrifice. Freud wrote in *Civilization and Its Discontents*: "The sense of guilt, the harshness of the super-ego, is ... the same thing as the severity of the conscience. It is the perception which the ego has of being watched over in this way ... the need for punishment, is an instinctual manifestation on the part of the ego, which has become masochistic under the influence of a sadistic super-ego" (Freud 1961, p. 100). The Christian superego commandment to love your neighbor as yourself is psychoanalytically "impossible to fulfil", which leads the therapist "for therapeutic purposes, to oppose the super-ego, and we endeavor to lower its demands" (Freud 1961, pp. 107–8). Lacan called the superego's law of sacrifice a "dark God": "If the superego always demands more sacrifice, more work, this is because the ideal it sets in front of the subject is kept aloft by a loss that the subject is unable to put behind him. The superego attempts to mask the loss of the Other by posing as witness or reminder of that absolute satisfaction which can no longer be ours" (Copjec 2002, p. 46). Psychoanalysis is determined to expose the cruelty and otherness of this sadistic superego and to keep its distance from it.

In his study of such a "dark God" Lacan was greatly inspired by the Pauline dialectics between Law and desire: "The relationship between the Thing [i.e., sin] and the Law could not be better defined than in these terms … It is only because of the Law that sin … takes on an excessive, hyperbolic character. Freud's discovery—the ethics of psychoanalysis—does it leave us clinging to that dialectic?" (Lacan 1992, pp. 83–84). Lacanian psychoanalysis is an affirmation that there is a way to relate to the Thing "somewhere beyond the law", which, in Žižek's commentary, is "the possibility of a relationship that avoids the pitfalls of the superego inculpation that accounts for the 'morbid' enjoyment of sin" (Žižek 1999, p. 153). Lacan's maxim, "don't give way on your desire", no longer refers to the desire involved in the morbid dialectic with Law, but to desire as equivalent to fulfilling your ethical duty. Entangled in the Pauline mutual involvement between Law and desire is the paradox of the superego, which enjoys pleasure in feeling guilty by producing "This perverse universe in which the ascetic who flagellates himself on behalf of the Law enjoys more intensely than the person who takes innocent pleasure in earthly delights—is what St Paul designates as 'the way of the Flesh' as opposed to 'the way of the Spirit': 'Flesh' is not flesh as opposed to the Law, but flesh as an excessive self-torturing, mortifying morbid fascination *begotten by the Law*" (Žižek 1999, p. 150). This morbid self-sacrifice was what repelled Nietzsche, who wrote: "Finally: what is left to be sacrificed? Did not one have to sacrifice everything comforting, holy, healing, all hope, all faith in a concealed harmony, in a future bliss and justice? Did one not to have to sacrifice God himself … ? To sacrifice God for nothingness—this paradoxical mystery of the ultimate act of cruelty was reserved for the generation which is even now arising" (quoted in Keenan 2005, p. 60).

What one learns from Paul is that a true Christian life is not based on the superegoic prohibition and struggle for self-sacrifice, but on the affirmative prospect of *agape*. If sacrifice has a transcendental intention towards the superegoic Other who inaugurates the cycle of Law and desire, "*Agape*–as the sacrifice *of* the sacrifice of one's 'pathological' sinful desire to transgress the Law … –is what St. Paul calls 'dying to the law'" (Keenan 2005, p. 130). Thus, Paul does not preach an economy of sacrifice that pays, in which one suffers in this world to get a reward in the other, as Nietzsche accused him; his agape is spontaneous work without expecting a reward, sacrifice that is not for a Cause but *for nothing*, after one has experienced, in Lacan's terms, "symbolic death" or "subjective destitution".

4. Yoyes' Breakthrough: Unmasking the Forced Choice

There was a militant in ETA who broke with the traditional male model of heroism—Yoyes, the nom-de-guerre of María Dolores González Cataraín, one of the teenage girls in the organization's early 1970s underground. She was forced into exile in 1974. By 1978, she held one of ETA's highest leadership positions. However, the following year she decided to abandon the armed organization and start a new life in Mexico, where she studied sociology and, in 1982, had a son. She returned to Paris in 1985 and then settled in Donostia-San Sebastián with her son and her partner. On 10 September 1986, while visiting her town during the Basque fiestas, she was shot and killed as a traitor by her former comrades in front of her three-year-old son.

With her decision to challenge ETA, Yoyes rejected the symbolic order of her own former militant identity as a condition of her autonomous ethics. Her alienation began with her realization of the machismo behind her ETA comrades' attitudes. She wrote in her diary that introducing feminist perspectives into the underground organization was a "most urgent task", adding, "What should I do for these men to understand and fully assume that women's liberation is a revolutionary priority?" Not only does she reject the machismo of her comrades, she is also afraid that it might infect her as well: "I don't want to become the woman who is accepted because men consider her in some way macho" (Garmendia Lasa et al. 1987, p. 57). When the organization repeatedly tried to lure her back to armed activism, she described their efforts as something akin to those of "a spurned husband abandoned by his wife" (Garmendia Lasa et al. 1987, p. 166). In her writings, Yoyes describes the radical changes she experienced in the coordinates of her subjectivity. She had the unique courage to openly take the position that "in the modern ethical constellation ... one *suspends this exception of the Thing*: one bears witness to one's fidelity to the Thing by *sacrificing (also) the Thing itself*" (Žižek 2000, p. 154). In both her surrender to and then her overcoming of the ethics of martyrdom, Yoyes became ETA's most consequential member. She embodied the Kierkegaardian paradox of "being a martyr without the martyrdom associated with being a martyr" (Copjec 1999, p. 258).

Yoyes persevered in her new freedom until she was murdered. What Copjec wrote about Antigone applies to her: "Perseverance does not consist in the repetition of a 'pattern of behavior', but of the performance, in the face of enormous obstacles, of a creative act, and it results not in the preservation of the very core of her being—however wayward or perverse—but of its complete overturning. Antigone's perseverance is not indicated by her remaining rigidly the same, but by her *metamorphosis* at the moment of her encounter with the event of her brother's death and Creon's refusal to allow his burial" (Copjec 1999, p. 258). ETA's refusal to allow Yoyes' own desire to have a child and an ordinary family life turned her into an unyielding rebel, this time not in defiance of Spanish rule, but against her former comrades. She persevered by keeping the faith, not to a nationalist allegiance, but to an inner ethical core. Yoyes' drama was, as Butler wrote of Antigone, "a conflict internal to and constitutive of the operation of desire and, in particular, ethical desire" (Butler 2000, p. 47). By her decision to oppose ETA, Yoyes, who writes of a feeling of "entombment", made of herself, like Antigone, a figure "between two deaths". Yoyes' decision to disobey ETA shows her determination not to compromise her desire, even if this implied death. But, in the case of both Antigone and Yoyes, "Her 'criminal desire' is *not* the sacrifice for a cause (and therefore a desire mediated by one's alienation in/by the symbolic order), but rather the sacrifice of the sacrifice, which is a separation from the symbolic order" (Keenan 2005, p. 116).

It is hard to overestimate the breakthrough effected by Yoyes. Not only had she given herself entirely to the "terrorist" cause of Basque independence for a decade, but she also ended up sacrificing the Cause/Exception itself of her own fight. If Abraham had been willing to sacrifice his son for the sake of the big Other, Yoyes would not. Yoyes would become the first ETA militant to show that the glorification of the sacrificial hero was a *masculine* affair. "I don't like the business of heroism", she wrote in her diary. Begoña Aretxaga summed up best the conundrum posed by Yoyes to ETA: "Hero, traitor, martyr—Yoyes was everything that, from the cultural premises embedded in nationalist practice, a woman could not be. Moreover, Yoyes was a mother. In the nationalist context, the models of hero, traitor or martyr and the model of the mother are mutually exclusive. It is precisely, I believe, the synthesis of these models in the person of Yoyes which made her 'treason' much more unbearable than that of other ex-militants" (Aretxaga 2005, p. 158).

Yoyes not only lived for a decade by the axiom "Freedom or Death", but she also forced an evolution in that ultimate alternative by unmasking that ETA had corrupted the empowering revolutionary dilemma into a *forced choice*—the kind of choice faced by the mugger's alienating dilemma, "Your money or your life", where the alternative resides entirely in the realm of the Other. The radicalness of Yoyes' act consisted precisely in having transformed the understanding and reality of "freedom" and "death" in the revolutionary dilemma. From her beginning with ETA, "death" had intersected with "freedom" in the revolutionary domain, but later, for Yoyes, both terms collided in her own gendered being.

Lacan paid closed attention to the structure of such a "forced choice". He wrote: "*Your freedom or your life*! If he chooses freedom, he loses both immediately—if he chooses life, he has life deprived of freedom" (Lacan 1998, p. 212). Only in theory can you choose one of the alternatives, in reality if you want to preserve your freedom of choice you can only choose one of the two, for in "*freedom or death*!, the only proof of freedom that you can have in the conditions laid out before you is precisely to choose death, for there, you know that you have freedom of choice" (Lacan 1998, p. 213). The structure of the "forced choice", by which one "chooses" what is already given, a choice in which only one alternative is valid, when the subject is forced to make the "empty gesture" of choosing as his own what is already there, "*is* the symbolization of the Real, the inscription of the Real into the symbolic order" (Keenan 2005, p. 112).

As Yoyes resexualized her life and rejected ETA's forced choice, the fusion of love and death took on a different dynamic. She wrote in her diary: "'To be ready to give your life' cannot mean 'to be ready to surrender your life to the enemy,' they are two totally different things, I would say they are opposed" (Garmendia Lasa et al. 1987, p. 68). In the revolutionary alternative, the meaning of "death" could be read literally in biological terms. But for the post-ETA Yoyes, the meaning of death is rather the psychoanalytic notion of the "death drive", which is not opposed to the "life drive", but rather both drives emerge from the same erotic core. When Yoyes decides to resexualize her life by giving priority to having her son, the fusion of love and death takes on a different dynamic. In her new life, death will keep intersecting her subjectivity, but only in terms of the "death drive", not biological sacrificial death. The ethical act by which Yoyes changes the coordinates of the sacrificial politics of ETA is summed up in the transformation of the "freedom or death" alternative, which she rescues from the mugger's forced choice under the threat of physical death to a death drive that is fully eroticized in a corporeal manner, culminating in an intersection that allows for a free choice to be made by the ethical actor.

In psychoanalysis, the forced choice is tied of to the formation of the big Other. It took the historic rupture of Yoyes to see the link between ETA as the big Other of Basque politics and to lead others to rebel against the turning of its revolutionary alternative into a forced choice. Yoyes' breakthrough meant that she came to see the unconscious link between the political superego, male symbolic castration, and the need for sacrifice. "Symbolic castration" is the psychoanalytic name for "the loss of the Real" upon the emergence of the subject into the symbolic order, the sacrifice of the incestuous Thing at the origin of individual consciousness; it is also the name for the price one has to pay when one is acting not in one's own name, but in the name of a superior Other that one embodies. Lacan described with the distinction between "feminine" and "masculine" modes of subjectivity regarding the "phallic function". He concluded that there is, on the female side, a fundamental undecidability, referred to as "not-all" (not all of her is subject to the phallic rule), whereas, on the male side, all of man is subject to such a rule. The "feminine" subjectivity relies on an ontological definition of being plural and partial; woman does not form an "all"; "she is not susceptible to the threat of castration" (Copjec 2002, p. 35). Lacan's conclusion was that the castrated one is not the woman, as Freud thought, but it is the man who is completely dependent on the phallic signifier and therefore more frequently subjected to symbolic castration. The prohibition of the Father, on the other hand, inaugurates the domain of the superego—the internalization of ideals fashioned by society. The superego is, for Lacan, "a correlate of castration" (Lacan 1999, p. 7). In the original scenario of castration, the boy, not the girl, is subjected to the father's prohibition. Castration is enacted for boys as a prohibition that comes from a "beyond"—the law that inaugurates the superego. It is this cruel superego that is always thirsty for sacrifice and that affects masculinity in particular. ("Feminine" and "masculine" are not substantive gendered realities nor are they trapped in any binary logic, rather they involve two subjective modalities).

The historic rupture brought about by Yoyes consisted of traversing through the unconscious links between symbolic male castration, its political superego, and the need for sacrifice. In ETA, Yoyes had become "the man" by imposing a different subjectivity. She was the one who showed her comrades, who had defined themselves as Guernica's victims, the transposition by which they had turned into executioners themselves. Like Antigone, Yoyes "is *destined to overturn her fate through her act*" (Copjec 2002, p. 45). The same ETA militants who assassinated her would soon embrace Yoyes' positions and call for an end to sacrificial politics. After the Yoyes event, ETA could no longer be the same. Yoyes had sacrificed sacrifice.

5. From Antigone to Sygne: Yoyes at the Window

Antigone's fate took place in a context of tyranny, one in which the individual lacks the possibility of choosing because the master has chosen for her. Etxebarrieta's commitment found its fate in a situation of modern military dictatorship, and so did Yoyes when she made her decision to join the underground in the early 1970s. But when military tyranny was replaced by democracy in Spain after Franco's death in 1975, thanks to Yoyes, the coordinates of the armed struggle changed. In the ensuing debate within ETA as to how to proceed, she found herself alienated from her organization. Until then, she had embodied Antigone's unflinching rejection of the Spanish dictatorship. But with the change in the political context, should Antigone continue to be her unique model?

The fate of Antigone has been contrasted with that of another modern heroine, Paul Claudel's Sygne de Coûfontaine in *L'Otage*. Sygne's fate occurs in modern France during the revolutionary period that haunts the Ancient Regime. Sygne is forced to choose between marrying the executioner of her family, the loathsome Turelure, and making him the Lord of Coûfontaine, or being arrested in the company of Georges (the cousin to whom she has sworn eternal love) and the Pope (who is hiding at home after escaping from French captivity). After talking to her confessor, the devout Sygne marries Turelure for the sake of saving her noble House and the Pope. She bears him a son. Later on, as her cousin is about to fire a bullet at Turelure, Sygne jumps to shield her husband and receives the fatal shot; Turelure asks from Sygne a sign to give some meaning to her suicidal act of saving his life, not out of love for him but just to save the family name; she refuses a final pardon and reconciliation, her only expression a compulsive tic in her lips signaling a "no". So, in the end, Sygne sacrifices even her own religious principles of love and forgiveness, for which, until then, she had been willing to sacrifice everything else. Several commentators, following Lacan, see in this sacrifice *the exception* of what can be sacrificed as a paradigm of the true ethical act. While Antigone transgressed the laws of the city and died a sublime heroine, Sygne dies in abjection with no cause and no pride left.

Alenka Zupancic writes about Sygne's choice: "terror presents itself in those situations where the only way you can choose A is by choosing its negation, non-A; the only way the subject can stay true to her Cause is by betraying it, *by sacrificing to it the very thing which drives her to make this sacrifice*. It is this paradoxical logic which allows subjectivation to coincide here with the 'destitution' of the subject" (Zupancic 2000, p. 216). Something similar applies to Yoyes: after experiencing that the revolutionary alternative "freedom or death!" had turned into a forced choice, she can only stay true to the Cause of freedom by betraying its initial revolutionary slogan, by sacrificing to freedom the very revolutionary ideal that drove her to make the sacrifice of her life.

A paradigmatic case of the terror of forced choice is William Styron's novel, made into the film *Sophie's Choice*, in which Sophie, as she arrives in Auschwitz with her two children, is forced by the German officer to choose who of the two children will be saved and who will go to the gas chamber: "Sophie loses more than a child ... she must sacrifice something more than anything she *has* ... she has to sacrifice what she *is*, her being which determines her beyond life and death" (Zupancic 2000, pp. 214–15).

Signe's final "no" before dying signals that she did not give up her desire, for "it is characteristic of the logic of desire itself to have as its ultimate horizon the sacrifice of the very thing in the name of which Sygne is ready to sacrifice everything" (Zupancic 2000, p. 229). This was the negation of negation, the multiple sacrifice of sacrifice. That final negation was only possible after her initial choice; the confessor did not ask her to love Turelure, only to marry him to save the Pope in an act that can be seen as religious but that did not prevent her from not giving up her desire beyond desire. From the time of her forced choice, Sygne surrenders the life she *has* and the honor she *is*, but at her last breath she still refuses to disappear and denies any divine sublimity of a final reconciliation. This is the moment of "pure desire", which "can be defined as the moment when the only way for the subject not to give up on her desire is to sacrifice the very Cause of her desire, its absolute condition ... pure desire can be defined as the moment at which desire is forced to say for its own Cause (for its absolute condition): 'That's not It'. This means that the moment of pure desire is, paradoxically, the very moment at which desire loses the foundation of its purity" (Zupancic 2000, p. 244). After a decade of underground militancy, Yoyes found herself protesting at her comrades in arms: "That's not it!"

The trouble with Antigone's type of ethics of desire is the role fantasy plays in it. Since "desire is nothing but that which introduces into the subject's universe an incommensurable or infinite measure", from such a perspective, "to realize one's desire means to realize, to 'measure' the infinite, the infinite measure" (Zupancic 2000, p. 251). The infiniteness of desire is of a negative magnitude in that it has no end—which is Hegel's "bad infinite". This type of desire lacks any temporal dimension, it is ruled ultimately by fantasy: it has no capacity to frame the fantasy from which one can contemplate the spectacle of one's actions. Despite all her sublime beauty, and even if "it might seem paradoxical", we could "link the figure of Antigone to the 'logic of fantasy' in this way" (Zupancic 2000, p. 253). She is unable to experience any feeling of the sublime in her suicidal action because she has no frame to impose on her fantasy, she is inside it—as Lacan put it, "from Antigone's point of view life ... can only be lived or thought about, from the place of that limit where her life is already lost, where she is already on the other side" (Lacan 1992, p. 280). In this regard, "the ethics of desire *is* the ethics of fantasy ... : we cannot deny all ethical dignity to someone who is ready to die (and to kill) in order to realize his or her fantasy" (Zupancic 2000, p. 254). We call these "anachronistic" people terrorists, fundamentalists, and madmen. We all have our fantasies but prefer not to realize them.

Etxebarrieta, not Yoyes, was Antigone. The words of protest by the Chorus against Antigone, in Žižek's play *Antigone*, apply to Etxebarrieta: "the greatest wisdom is to know when this very fidelity [to what can and cannot be said] compels us to break our word, even if this word is the highest immemorial law. This is where you went wrong, Antigone. In sacrificing everything for your law, you lost this law itself". Antigone replies: "I just stood for justice, whatever the costs. How can this be wrong?" Chorus: "We see how dedicated you are to your Cause, ready to sacrifice everything for it. But wisdom tells us that, sometimes, when you forsake everything for your Cause, what you lose is the Cause itself, so all your sacrifices were in vain, for nothing. Then you end up not as a noble hero but as an abject whose place is neither with the living nor with the dead, but in the uncanny in-between where monsters abide that our mind cannot even contemplate" (Žižek 2016, pp. 23–24).

Yoyes' diaries were published with the title "Yoyes from her window", displaying a photo of her at a window on the cover: the window was the metonym for her attempt at creating a new life for herself by putting a frame on her fantasy and desire. She was at the window watching the equivalent of a horror movie unfolding in front of her eyes after she returned with her son and partner to a civilian life. How could she show her comrades and the Basque public that what she needed was to go beyond the ethics of fantasy? Framing was necessary to see the change in the status of knowledge throughout the history of ETA. If Etxebarrieta's initial ETA, as a blind Oedipus, did not know where his choice for martyrdom would lead the organization, after democracy and years of armed militancy, Yoyes, like Sygne, was like "an Oedipus who knows" (Zupancic 2000, p. 256). A change in the symbolic constellations had taken place: not only was the big Other of Franco dead, but, after the transition to democracy during the late 1970s, ETA itself, the big Other of the Basque resistance to fascism, had lost its raison d'être and turned itself, for most Basques, into an anachronistic remnant of Francoism. Yoyes' historic role was to show that ETA, as the big Other, was dead and that the symbolic debt owed to the Cause embodied by ETA had lost its unconditional value. There was no sublime heroism a la Antigone for Yoyes; she was, instead, like Sygne, who sacrificed even the ground for her Cause.

6. Conclusion: The Sacrifice of Sacrifice

The theme of the sacrifice of sacrifice has been studied among others, in the wake of the work of Hegel and Nietzsche, by Lacan, Derrida, and Žižek, and by Kierkegaard,

Bataille, Blanchot, Levinas, Kristeva, and Irigarai. Following the Hegelian logic of "the negation of negation", the sacrifice of sacrifice is, for these authors, an authentic ethical act. It is a topic that has become pivotal for any assessment of historical processes, such as the one I am attempting here. There was nothing harder and more consequential for many of my own post-war Basque generation than such a sacrifice of sacrifice in the various domains of religion, politics, sex, or militant culture in general. As many were forced to sacrifice religion in order to keep its ethical core alive, the Basque radical Left also finally found the courage to "sacrifice" ETA for the sake of keeping their fidelity to the political project that gave birth to it.

The Hegelian dialectical system has been caricaturized as a progression from thesis and antithesis to synthesis, but the second negation is not a mere synthesis of the opposites but rather a more radical negation that negates the first symbolic position. There is no simple progression or succession between the two negations; according to Hegel, "the very initial immediacy is always-already 'posited' retroactively, so that its emergence *coincides with its loss*" (Žižek 2002, p. 167; also v Žižek 2012, pp. 292–304). In other words, "negation is itself negated; sacrifice is itself sacrificed. Essence 'is' *nothing but* redoubled reflection, *nothing but* radical negativity, *nothing but* radical sacrifice that cannot not be (dis)embodied in 'appearance'" (Keenan 2005, p. 106). In the move from the first negation to the second "negation of negation", there is a change from the objective to the subjective—in the second stage, the subject, who sees the results of his own position, includes himself in the process. Objectively, the "crucifixión" marks the death of God and there can be no more extreme negation, but, in its double negation, it turns into the space of subjective freedom—Christ's death turns into "the death of death".

The Hegelian logic by which the subject posits his/her own ground in a redoubled negation is echoed in the basic psychoanalytic experience of knowledge through misrecognition, where truth is produced through the structural illusion of transference. The subject has to first be deceived by the call of the Other before recognizing its inexistence, what leads to the experience of "subjective destitution". Thus, Job had nothing to complain about, what happens to him is nothing exceptional, there is no secret meaning to it, unless the secret is God's own impotence. What self-relating negativity demands from Job is not only to accept the utter despair of the complete loss that has befallen him, but also to get rid of the loss itself ("the loss of the loss") in the sense of not expecting to regain any of the losses, but finding "a radical void after losing the very coordinates which made the loss meaningful" (Žižek 2012, p. 478).

In the case of ETA, it was the sacrificial model of the Crucified that its own public perceived from the very beginning (Zulaika 1988). This sacrificial duty inherited from the Abrahamic religious traditions is the one described by Derrida as "the gift of Death" (Derrida 1995). "Greater love has no man than this, that he lay down his life for a friend" (John 15:13). From Plato's Socrates to Heidegger, at the very heart of Western thought resides the idea that willingness to surrender your life for someone else's sake is a supreme expression of love and freedom, the ultimate triumph of life. Derrida sees the history of the West grounded on such a measureless principle, including the commandment to give and take human life as something imposed by modern states on their citzens. This is, in short, what Kierkegaard reads into the story of Abraham: the ultimate duty and aporia of responsibility, as well as the ultimate mockery of ethics, is human sacrifice. Derrida insists that the sacrifice of Isaac cannot be erased from the tradition of the three Abrahamic religions. This is the Christian *mysterium tremendum*, Kierkegaard's "fear and trembling" when confronted with the experience of life as sacrifice. What does it mean to "give yourself death", to be responsible for it, to accept the gift of death for another as Socrates, Christ and so many others did? Derrida debates these issues while he examines the founding position of sacrifice has in the thought of Kierkegaard, Heidegger and other thinkers. One can die for someone but not instead of someone else. In such a philosophical tradition that begins partly with Kant and Hegel, a thought that "repeats" the possibility of religion without religion, the logic of sacrifice becomes concrete in that all death is in the end is a donation; thus, death also brings life, a notion confirmed by world ethnography.

But there is one lesson Derrida cannot avoid drawing from Abraham: "[W]hat does Abraham teach us, in his approach to sacrifice?" Derrida asks before replying: "That far from ensuring responsibility, the generality of ethics incites to irresponsibility" (Derrida 1995, p. 61). For Abraham, writes Kierkegaard, "the ethical is the temptation" (Kierkegaard 1941, p. 115). He overrides his ethical responsibility towards his son by feeling bound to another absolute responsibility, which is inconceivable, and about which Abraham cannot speak. The absurdity of using the notions of responsibility and duty to justify arbitrary murder turn the story of Abraham towards the conceptual limits of paradox, scandal and aporia, for "Abraham is faithful to God only in his absolute treachery" (Derrida 1995, p. 68). While the religious expression of his action is *sacrifice*, the ethical expression is no other than *murder*. "Abraham is therefore at no instant a tragic hero", Kierkegaard concluded, "but something quite different, either a murderer or a believer" (Kierkegaard 1941, p. 67). The problem with this Kierkegaardian logic in the domain of politics is that it is a bottomless abyss that would never reach the end of murder.

For Nietzsche, this sacrificial *hubris* was the "stroke of genius called Christianity". This is an economy, concludes Derrida, that is taken "to its excess in the sacrifice of Christ for love of the debtor; it involves the same economy of sacrifice, the same sacrifice of sacrifice" (Derrida 1995, p. 114). One must sacrifice calculated sacrifice (the one looking for reward or recognition) to preserve true sacrifice, as such. This leads, ultimately, to the double bind of religion, in that it "*both requires and excludes sacrifice*" (Derrida quoted in Keenan 2005, p. 158), in that it requires a sacrifice of sacrifice. Keenan sums up Derrida's position towards religion thus: "Bearing witness to the infinite transcendence of what is worth more than life [which] requires, therefore, not only a sacrifice in the name of transcendence, but also a sacrificing of transcendence ... [which] is a sacrificing of that in the name of which one sacrifices, which is a sacrificing of the very reason of sacrifice, insofar as sacrifice involves a transcendental intention. A sacrificing of transcendence is, therefore, a sacrificing of sacrifice" (Keenan 2005, p. 158).

Basque nationalism is no exception to the psychoanalytical truism that loss is constitutive of the subject; what demands perennial sacrifice is the effort to regain the lost liberties, laws and sovereignty of the past. ETA was fueled by such centuries-old loss, tragically reenacted most recently in Guernica. In the militant actor's subjective economy, sacrificial exchange for the freedom of the country was nothing but the dutiful thing to do. It was always doubtful that this sacrificial exchange would achieve its ultimate goal of erasing the original loss. But even if this was not the case, there was a basic factor that made the sacrifice necessary, namely to ascertain the existence of some Other out there. Suffering and defeat had a purpose and an explanation with ETA; without it, the world was a blind piece of machinery ruled by chance. "The sacrifice signifies that, in the object of our desires, we try to find evidence for the presence of the desire of this Other that I call here *the dark God*", wrote Lacan (1998, p. 275). Beyond affirming the existence of the big Other, the subject offers his/her sacrifice "to fill in the lack *in the Other*, to sustain the appearance of the Other's omnipotence or, at least, consistency" (Žižek 2001, p. 70). Sacrificing sacrifice meant, for those who had formed their basic political identity around ETA, that the world became meaningless as they had to give up what granted consistency to it.

The lesson to be learned by ETA's generation is the one that derives from psychoanalysis, whose aim "is not to enable the subject to assume the necessary sacrifice (to 'accept symbolic castration', etc.) ... but to *resist* the terrible attraction of sacrifice—attraction which, of course, is none other than that of the superego. Sacrifice is ultimately the gesture by means of which we aim at compensating the guilt imposed by the impossible superego injunction" (Žižek 2001, p. 74). Exorcising the passion for sacrifice has been the hardest subjective task for many of the ETA generation.

Funding: This research received no external funding.

Institutional Review Board Statement: Not applicable.

Informed Consent Statement: Not applicable.

Acknowledgments: This paper has benefitted from the insightful comments of Josetxo Beriain and an anonymous reviewer.

Conflicts of Interest: The author declares no conflict of interest.

References

Alvarez Enparantza, Jose Luis. 1977. *Leturia-Ren Egunkari Ezkutua*, 2nd ed. Durango: Leopoldo Zugaza.
Alvarez Enparantza, Jose Luis. 1985. Unamuno eragille. In *Gertakarien Lekuko*. Donostia: Haranburu, pp. 71–80.
Aretxaga, Begoña. 2005. *States of Terror: Begoña Aretxaga's Essays*. Edited by Joseba Zulaika. Reno: Center for Basque Studies.
Bloch, Maurice. 1992. *Prey Into Hunter: The Politics of Religious Experience*. Cambridge: Cambridge University Press.
Bloch, Maurice. 2013. *In and out of Each Other's Bodies: Theory of Mind, Evolution, Truth, and the Nature of the Social*. London: Paradigm Publishers.
Bullen, Margaret. 2003. *Basque Gender Studies*. Reno: Center for Basque Studies.
Butler, Judith. 2000. *Antigone's Claim: Kinship Between Life and Death*. New York: Columbia University Press.
Copjec, Joan. 1999. The Tomb of Perseverance: On Antigone. In *Giving Ground: The Politics of Propinquity*. Edited by Joan Copjec and Michael Sorkin. London and New York: Verso, pp. 233–66.
Copjec, Joan. 2002. *Imagine There Is No Woman: Ethics and Sublimation*. Cambridge: MIT Press.
Del Valle, Teresa, Joxemartin Apalategi, Begoña Arregui, Begoña Aretxaga, Isabel Babace, Mari C. Díez, Carmen Larrañaga, Amparo Oiarzabal, Carmen Pérez, and Itziar Zuriarrain. 1985. *Mujer Vasca: Imagen y Realidad*. Barcelona: Anthropos.
Deleuze, Gilles. 1989. Coldness and Cruelty. In *Gilles Deleuze and Leopold von-Sacher Masoch, Masochism*. New York: Zone Books.
Derrida, Jacques. 1995. *The Gift of Death*. Chicago: University of Chicago Press.
Etxebarrieta, Jose Antonio. 1999. *Los Vientos Favorables: Euskal Herria 1839–1959*. Edited by José María Lorenzo Espinosa and Mikel Zabala. Tafalla: Txlaparta.
Evans-Pritchard, Edward. 1963. The Divine Kingship of the Shilluk of the Nilotic Sudan. In *Essays in Social Anthropology*. New York: Free Press of Glencoe.
Frazer, James George. 1963. *The Golden Bough: A Study in Magic and Religion*. New York: Macmillan.
Freud, Sigmund. 1950. *Totem and Taboo*. New York: Norton & Company.
Freud, Sigmund. 1961. *Civilization and Its Discontents*. New York: Norton & Company.
Garff, Joakim. 2000. *Soren Kierkegaard: A Biography*. Princeton: Princeton University Press.
Garmendia Lasa, Elixabete, Gloria Gonzalez Katarain, Ana Gonzalez Katarain, Juli Garmendia Lasa, and Juanjo Dorronsoro. 1987. *Yoyes Desde su Ventana*. Pamplona: Garrasi.
Girard, René. 1977. *Violence and the Sacred*. Baltimore: Johns Hopkins University Press.
Keenan, Dennis King. 2005. *The Question of Sacrifice*. Bloomington: Indiana University Press.
Kierkegaard, Soren. 1941. *Fear and Trembling and the Sickness unto Death*. Princeton: Princeton University Press.
Lacan, Jacques. 1992. The Ethics of Psychoanalysis 1959–1960. In *The Seminar of Jacques Lacan*. Edited by Jacques-Alain Miller. New York: W. W. Norton & Company, vol. 7.
Lacan, Jacques. 1998. The Four Fundamental Concepts of Psychoanalysis. In *The Seminar of Jacques Lacan*. Edited by Jacques-Alain Miller. New York: W. W. Norton & Company, vol. 11.
Lacan, Jacques. 1999. On Feminine Sexuality: The Limits of Love and Knowledge, 1972–1973. In *The Seminar of Jacques Lacan*. New York: W. W. Norton & Company, vol. 20.
Lacan, Jacques. 2013. *On the Names-of-the-Father*. Translated by Bruce Fink. Malden: Polity.
Lorenzo Espinosa, José María. 1994. *Txabi Etxebarrieta: Armado de Palabra y Obra*. Tafalla: Txalaparta.
Lorenzo Espinosa, José María. 1996. *Txabi Etxebarrieta: Poesía y Otros Escritos 1961–1967*. Tafalla: Txalaparta.
Miller, Jacques-Alain. 2009. The Prisons of *Jouissance*. *Lacanian Ink* 33: 36–55. Available online: https://www.lacan.com/lacinkXXXIII3.html (accessed on 10 January 2021).
Nietzsche, Friedrich. 1967. *On the Genealogy of Morals*. Translated by Walter Kaufmann, and R. J. Hollingdale. New York: Random House.
Onaindia, Mario. 2001. *El precio de la libertad: Memorias (1948–1977)*. Madrid: Espasa.
Ortiz-Osés, Andrés. 1980. *El matriarcalismo Vasco*. Bilbao: Publicaciones de la Universidad de Deusto.
Rappaport, Roy. 1979. *Ecology, Meaning, and Religion*. Richmond: North Atlantic Books.
Roudinesco, Élisabeth. 2016. *Freud: In His Time and Ours*. Translated by Catherine Porter. Cambridge: Harvard University Press.
Sahlins, Marshall, and David Graeber. 2017. *On Kings*. Chicago: Hau Books.
Unamuno, Miguel de. 1972. *The Tragic Sense of Life in Men and Nations*. Translated by Anthony Kerrigan. Princeton: Princeton University Press.
Uriarte, Teo. 2005. *Mirando Atrás: De las Filas de ETA a las Listas del PSE*. Barcelona: Ediciones B.

Zambrano, María. 1995. *Las Palabras del Regreso*. Edited by Mercedes Gómez Blesa. Salamanca: Amarú Ediciones.
Žižek, Slavoj. 1999. *The Ticklish Subject: The Absent Center of Political Ontology*. London and New York: Verso.
Žižek, Slavoj. 2000. *The Fragile Absolute*. London: Verso.
Žižek, Slavoj. 2001. *On Belief*. London and New York: Routledge.
Žižek, Slavoj. 2002. *For They Know Not What They Do: Enjoyment as a Political Factor*. London: Verso.
Žižek, Slavoj. 2008. *Violence: Six Sideways Reflections*. New York: Picador.
Žižek, Slavoj. 2012. *Less Than Nothing: Hegel and the Shadow of Dialectical Materialism*. London: Verso.
Žižek, Slavoj. 2016. *Antigone*. London: Bloomsbury.
Zulaika, Joseba. 1988. *Basque Violence: Metaphor and Sacrament*. Reno: University of Nevada Press.
Zulaika, Joseba. 2014. *That Old Bilbao Moon: The Passion and Resurrection of a City*. Reno: Center for Basque Studies Press.
Zupancic, Alenca. 2000. *Ethics of the Real*. London and New York: Verso.
Zupancic, Alenca. 2003. *The Shortest Shadow: Nietzsche's Philosophy of the Two*. Cambridge: MIT Press.

Article

Debt and Sacrifice: The Role of Scapegoats in the Economic Crises

Luis Enrique Alonso and Carlos J. Fernández Rodríguez *

Department of Sociology, Universidad Autónoma de Madrid, 28049 Madrid, Spain; luis.alonso@uam.es
* Correspondence: carlos.fernandez@uam.es

Abstract: Despite the process of secularization and modernization, in contemporary societies, the role of sacrifice is still relevant. One of the spaces where sacrifice actually performs a critical role is the realm of modern economy, particularly in the event of a financial crisis. Such crises represent situations defined by an outrageous symbolic violence in which social and economic relations experience drastic transformations, and their victims end up suffering personal bankruptcy, indebtedness, lower standards of living or poverty. Crises show the flagrant domination present in social relations: this is proven in the way crises evolve, when more and more social groups marred by a growing vulnerability are sacrificed to appease financial markets. Inspired by the theoretical framework of the French anthropologist René Girard, our intention is to explore how the hegemonic narrative about the crisis has been developed, highlighting its sacrificial aspects.

Keywords: violence; sacrifice; expropriation; crisis; financialization; capitalism

Citation: Alonso, Luis Enrique, and Carlos J. Fernández Rodríguez. 2021. Debt and Sacrifice: The Role of Scapegoats in the Economic Crises. *Religions* 12: 128. https://doi.org/10.3390/rel12020128

Academic Editors: Javier Gil-Gimeno; Josetxo Beriain and Celso Sánchez Capdeq

Received: 21 December 2020
Accepted: 14 February 2021
Published: 17 February 2021

Publisher's Note: MDPI stays neutral with regard to jurisdictional claims in published maps and institutional affiliations.

1. Introduction

Sacrifice performed an important role in agrarian societies and in their religious and cultural practices, acting as a gateway between the realms of the sacred and the profane (Mauss 1979) and being particularly relevant for soteriology (see, e.g., Eberhart and Schweitzer 2019). The concept of sacrifice has a religious origin and can be understood as a means of communication between the sacred and the profane, with the mediation of an immolation (Hubert and Mauss 1981). According to Hénaff (2002, 2012), in agrarian societies, there is a step in human evolution where humans are able to domesticate plants and animals, a capability regarded as taken from the spirits and gods. The capacity to breed plants and animals and consume them is interpreted as a gift from the deities, which needs to be returned in some form, in a logic of symbolic exchange that connects humans and non-humans, as well as physical forms of subsistence with symbolic representations of life. Therefore, sacrificial rituals basically emerge as an offering to the deities, in which humans return symbolically what they have taken from nature. The unilateral gift of grace is recognized with the sacrifice, with immolation making the offering irreversible. Gradually, this recognition goes beyond the animal world to involve the entire world around us, and the concept and practices of sacrifice experience a metamorphosis, adopting an increasingly metaphorical form (Hénaff 2012; see also Beriain 2017).

Despite the process of secularization and modernization, sacrifice is not absent from modernity. Different forms of sacrifice and self-sacrifice persist today, linked to religion and spirituality (e.g., self-immolations linked to new jihadism), but other relevant manifestations, yet more subtle, have emerged in secular life. Sacrifice still plays a key role in the realm of modern capitalist economy, where logics of exchange are as essential as in religion (Beriain 2017). Sociologists have often theorized the important links between the sphere of religion and economy (Weber 2005; Sombart 2001), where orientations and attitudes that once were spiritual evolved in secular society, becoming embedded in economic practices and behaviors[1]. The role of sacrifice in economy was remarkably explored by Georg Simmel, who, in his magnum opus, The Philosophy of Money (1978: original edition 1900), highlighted that the subjective condition of economic value does not arise from the pleasure of possessing an object, but rather from the desire derived from its non-possession and the sacrifices that the individual needs to endure in order to obtain such an object. Hence, the satisfaction of a desire has a price and involves a renunciation. The sacrifice involves not only committing to an ascetic lifestyle (giving up pleasure in order to save money), but also the striving in the context of market competition. The economic system is, according to Simmel, based on an abstraction: the mutuality of exchange between sacrifice and gain, where sacrifice becomes the condition of economic value as such (Simmel 1978, pp. 78–82). The individual accepts sacrifices in order to fulfil her desires at a later stage, leading to self-awareness and responsibility to oneself. Therefore, sacrifice is key to the contemporary economic order.

The role of sacrifice is particularly visible in certain events such as economic crises. In these scenarios, the sacrificial dynamics seem to function more in line with Durkheim's view, where sacrifice is understood as a ritual embodied in collective representations, creating belief in its efficacy and sacredness (Durkheim 1960). Thus, the resolution of a crisis involves a different type of logic, in which, in our view, others can be sacrificed for the benefit of the moral energy of the group. This perspective, in which the moral cohesion of society becomes crucial, pushes us to place the debate of economy and sacrifice within a different theoretical framework. In this paper, we aim to discuss the persistence of sacrifice in our secular age, and to do so, we will draw on René Girard's sometimes controversial theory of the scapegoat (see Girard 1986), which offers extremely useful insights into the relations among societies, the economy and the sacrificial symbolism. Girard's theory provides a narrative on crisis and sacrifice that certainly matches the logics displayed when societies confront economic cataclysms. The sharp downturn in economic activity generates uncertainty and leads to a political struggle to impose certain economic policies in order to resolve it. Such policies, defined by Peter Berger (1974) as a *calculus of pain* and designed by the political and economic powers, have a direct impact on citizens, affecting not just their standard of living but also existing economic and social relations themselves. Financial crises, therefore, can be best understood as phenomena that shape both the economy and social relations, as events in which some groups make gains and others lose in the struggle over the redistribution of resources. Sacrifice takes place when some groups, often unwillingly, are forced to live with less, in order to keep the social order in balance.

This process involves a factor that has generally received little attention yet should not be overlooked: crises usually unleash an extraordinary degree of economic and social violence, at least in symbolic terms (see Bourdieu 1991). This deserves more detailed consideration. The so-called victims of the crisis are citizens who, after a series of different contingencies in which large transformations in the world economy play an essential role, have lost their savings and/or income, and who, as a result, suffer personal bankruptcy, high debt, the loss of their livelihoods or a vicious spiral of poverty. This symbolic violence, as the result of a more profound social conflict, can be observed as a very real epiphenomenon which flagrantly exposes actually existing relations of social domination. Painful measures are necessary: we must send signals to the markets, make sacrifices. The latter idea, the notion of sacrifice, has in fact emerged as one of the leitmotifs of the Great Recession of 2008 in many European countries, one which refers to the need to sacrifice rights in order to send the markets signals that might calm their speculative fury. In such events, citizens' rights have somehow been chosen as the scapegoat in order, with the excuse of taming the violence of the markets (Aglietta and Orléan 1982), to reinforce the existing relations of social domination. The interesting analogy here with religion is that there are believers and interpreters (neoliberal economists, politicians) that portray

[1] According to Abend (2014), religious values can actually perform an important role as the moral background of business ethics.

a narrative of markets as sort of *Moloch* or divinity to whom no limits can be imposed in the current neoliberal framework, and whose cult—based on an endless accumulation of capital—demands, most of all, orthodoxy in following certain economic ideas and rules, no matter the costs[2]. This orthodoxy requires indeed certain sacrifices, even to the point where the economy would damage itself in an almost invoked eschatological whiff of the "End Times" (Peck 2013).

In this theoretical paper, we explore this notion of sacrifice in our secular age, highlighting its relevance in economic crises. Given the limited space, our article will focus on the theoretical discussion as well as some examples taken from the last financial crisis that the world witnessed: the Great Recession of 2008, where these logics of sacrifice were displayed in a quite dramatic way, particularly in Southern Europe. We have divided the rest of this paper into five sections. In Section 2, we argue that financial crises are, in essence, mechanisms for expropriating social wealth, highlighting the importance of the factors of control and punishment in the context of economic depression. In Sections 3 and 4, drawing above all on the theoretical framework put forward in Girard's work (see Girard 1977, 1986, 1987, 2001, 2003), we present an analysis of the crisis as sacrifice, linking these arguments to the important disciplinary role played by the crisis in societies. Next, we cite an example of how the sacrificial aspects of economic crises are deployed by examining the Great Recession in Southern Europe. In Section 5, we then discuss the notion of crisis as a disciplinary device designed to make people conform to global market rules. In Section 6, we end this paper with a brief conclusion.

2. Financial Crises as Mechanisms of Expropriation

It is remarkable to observe the scant importance attached to the mechanisms of social domination in the evolution of financial crises. Crisis as a justification for domination and mandatory sacrifice is a reality which, the more we feel its effects, the less we see it discussed in more conventional (official or academic) analyses of financial and social crises. When this dimension of crises is mentioned, it is treated as a mere collateral effect of economic decisions, presented as autonomous and sovereign, even if misguided. It is not worth examining here the most triumphant accounts of the superiority of deregulated financial flows, the trivialization of economic policies, the rational expectations of efficient capital markets and the infinite expansion of the markets in futures and derivatives, from Fama (1970) to countless neoliberal economists. In classic explanations of the crisis (e.g., Galbraith 1994; Montier 2000; Kahneman 2003; Kindleberger and Aliber 2005; Reinhart and Rogoff 2009; Akerlof and Shiller 2009), the references involve differing modes and degrees of eclipse of economic rationality, whether due to temporary madness, the imitation of high-risk or defensive behavior, undesired collective consequences of calculated individual actions or contagious miscalculation of risks. Above all, an almost universal blindness seems to lead to speculation and, from there, to fortunes and misfortunes The framework of power relations, the search for control and domination and the social victims of the collapse of institutions (for example, the non-investors) are mentioned, if at all, only in passing, as if they are anonymous extras in a drama in which they play no significant role.

However, these accounts are related neither to the social terrain in which they acquire their meaning, nor to the power strategies which explain them. The media and politicians frame a discourse where meteorological or physical metaphors for the crisis ("financial storms", "turbulence", "instability", "slump") conceal the actors involved, both those who suffer them and those that provoke them, but that somehow represents a state of nature where people cannot assert any significant influence, just like the weather. The crisis, therefore, can be understood as exploited in a discourse which constructs a specific social order, naturalizes power relations and makes all other alternatives impossible. Here, we

[2] It is interesting to notice how neoliberal economists have an almost religious belief that market (and private property) is the best way to ensure an individual's choice, while State is seen as dangerous. This Good (market and private property) vs. Evil (State and public sector) analogy has been extremely influential in countries such as the US, so it should not be surprising that some of the most extreme versions of neoliberalism have been actually supported by conservative faith-based groups (on this topic, see, e.g., Hackworth 2012).

follow Foucault (1980) in his argument that realities are presented as "natural", neutral and transparent, when, in truth, they are effective elements of the material production of knowledge and power. Hence, even when presented as such, the crisis is not a "natural" reality, but rather responds to the mechanisms and interests of the domination of meaning, which have operated throughout the history of capitalism. For all these reasons, every crisis symbolizes, in short, an ideology. The crisis mobilizes a certain symbolic violence (Bourdieu 1991, 2000) which gives rise to a situation of necessity and inevitability, generated by the unconscious adaptation of the objective and subjective structures. Exactly this makes it possible for private beliefs and interests to be embodied as if they were shared social objectives. At the same time, the crisis also permits the acceptance of dominant social categories (what is good, what is bad, what should be done, what should not be done, what is beneficial, what is unfavorable, what makes us better, what makes us worse, etc.), promoting attitudes of submission which are not perceived as such, but rather appear to be endowed with inherent legitimacy and absolute naturalness (Lordon 2010). This is linked to a specific state of exception (Agamben 2005) where a proliferation of emergencies leads to a new type of governmentality (Adey et al. 2015). The current crisis simply rounds off its disciplinary effect, as the private–mercantile powers impose their full weight, only constrained by the least institutionalized forms of social resistance.

In this respect, we would like to highlight two important issues. The first is that financial crises are essentially brought about by debt. As Adkins (2017, p. 450) claims, "debt has become the key mechanism through which economic and social existence is to be secured". Hence, it is a necessary but hazy corollary of other dominant (and certainly more attractive) concepts in contemporary mercantile discourse: leverage, financing, investment, mortgage, loan. The second is related to another element that usually goes unnoticed in day-to-day thinking about neoliberal capitalism: the violence (symbolic but, in certain occasions, very real) that capitalism unleashes on society in general and individuals in particular in order to force them to adapt to the new demands of neoliberal biopolitics and the rules of the current financial markets, rules which are happily sanctioned by the existing national and international mercantile, civil and criminal legal codes (see North et al. 2009).

Financial crises are, without doubt, directly related to the phenomenon of debt. In recent decades, debt, a societal–economic relation which seems to have accompanied humanity since the very beginning of time (Graeber 2011), has acquired a dominant role in generating economic growth. Indeed, capitalism can now be defined as debt-driven capitalism (Stockhammer 2009; Koch 2011; Poppe et al. 2016) in which debt plays an essential role in ensuring the very survival of the system (Lazzarato 2012). Financing consumers enables them to stimulate the capitalist economy enough for it to function; states are, to a large extent, reliant on public deficits to fund the various public services (Graeber 2011). Moreover, the consolidation of neoliberalism has also meant the hegemony of financial capital, whose hypostatization economists are loyal to, and supportive of. The new financialized regime has played a decisive role in the upsurge of a misleadingly named "popular capitalism", in which small investors and savers begin to participate, directly or indirectly, in channeling international monetary flows, simultaneously exposing their patrimonies to the vagaries of globalization. All this has had very serious implications for the question which concerns us here, that of debt.

In this respect, as noted by Graeber, the recent neoliberal cycle has also been distinguished by an authentic explosion in credit mechanisms created by the thriving financial sector as a means of making more and more profit. The proliferation of credit cards (American Express was founded in 1971, the same year that the United States abandoned the gold standard, thereby opening a new era of financial volatility) was accompanied by two other crucial developments with respect to debt. At the national level, in many countries, the first was the repeal, or at least watering down, of legislation against usury (one example is the US Monetary Control Act of 1980), permitting extremely high interest rates on various forms of personal loans and variable interest rates on mortgages (Aglietta and Orléan 1982).

This development has condemned many working- and middle-class families to a vicious cycle of indebtedness in order to cover their expenses, resulting in the need to live on credit and permanent indebtedness into a way of life recommended by the authorities and celebrated by economists and even clerics converted to the most rabid neoliberalism. Those people who do not need to get into debt or are unable to borrow money to fulfil their most basic obligations, such as paying for insurance or education, are either gods (the upper classes) or beasts (the excluded), to use an Aristotelian simile (Aristotle 1999). The second key development—in this case, international in scope—was the redefining of the mission of the International Monetary Fund (IMF) as a global institution responsible for working with international creditors to facilitate the Nation States' repayment of the debts that they contracted with the financial investors. This was accompanied by major legal reforms in other national and supranational bodies (Harvey 2005; Graeber 2011). These two developments have played a crucial role in the progressive financialization of the world and the progressive indebtedness of different economic agents (Harvey 2010).

The result of this tendency towards indebtedness is the growing importance of the financial sector in the economies of those countries which have been through the most intense processes of deregulation. The banks took advantage of liberalization to introduce new and sophisticated investment products to offer to their clients at the same time as they designed new credit formulae aimed at the middle and working classes. In this way, towards the middle of the first decade of the 21st century, the financial economy acquired a preposterous monetary value, which, moreover, was far higher than that of the manufacturing economy (Harvey 2005). This has had two consequences. First, as noted above, financial capital conditioned industrial capital, introducing a culture of short-term profit which led industrial companies to adopt strategies intended to satisfy their shareholders' immediate interests in profit, destroying the foundations of the Keynesian compromise[3] (Daguerre 2014). Second, the financial institutions' constant need for profit inevitably led them to make increasingly risky loans which, in turn, they insured and sold through complex financial products, giving rise to a spiral of interconnected and apparently limitless debt, for a time presented by governments and lobbies as a solid and harmonious model of economic growth (Krippner 2011).

However, this supposed virtuous circle of growth and debt sometimes crashes into its true limits, and it is then that the crisis becomes apparent in all its intensity. Here is where the theory of René Girard provides interesting insights. His concept of mimesis, developed throughout his work (see Girard 1977, 1987, 2001), states that we learn what to desire by copying the desires of others (mimetic desire), and this leads almost inevitably to conflict inside communities. Aglietta and Orléan (1982) applied this powerful idea to the way in which financial markets work. As these scholars have shown, monetary crises are, in essence, a sudden conflict between creditors and debtors, as the former try to enforce their rights in order to recoup their wealth from the latter, provoking a social conflict between the two—something which Girard argued is violent (see Girard 1977, 1986). The crisis, therefore, is generated by the inherent violence of the market economy (in which monetary violence has momentarily sublimated actual physical violence), which fuels a range of mimetic behavior in which each subject–individual imitates the rest, the other being simultaneously model and rival. According to Aglietta and Orléan (1982), this mimicry means that our economic decisions that imply purchases, credits, mortgages or personal loans are a mimetic response to the behavior of others: we imitate them and, at the same time, try to compete with them. In this light, financial crises are situations in which this sublimated violence is unleashed with the greatest force, since finances are also the field in which these relations of mimicry are strongest. Speculation is a classic case, breaking Walras' laws of price setting and generating a desire to hoard that mimics other speculators' practices and creates rivalries (see Galbraith 1994; Kindleberger and Aliber

3 The Keynesian/Fordist compromise was the dominant feature of post-war economics, critical for the development of the so-called welfare state. It implied a compromise between capital and labor that guaranteed economic stability and workers' participation in exchange for higher wages and increased social protection (see Alonso and Lucio 2006; Daguerre 2014).

2005). In neoclassical theory, the free market appears to consecrate individualism, while in fact generating a dread of difference, with the result that in most of our decisions, we end up imitating others. From a Girardian perspective, this could explain the rise of speculative bubbles of the type currently affecting the global financial markets, the behavior of senior executives and of a certain type of middle class, as well as, and above all, the collective desires suddenly aroused in situations of uncertainty which drive creditors to try to recover their investments even if this pushes their debtors to the brink of collapse.[4]

For Aglietta and Orléan (1982), therefore, the crisis is the moment when the economic agents try to satisfy their desires by asserting the rights which monetary sovereignty grants them, but find themselves facing a breakdown of legitimacy in the economic world (with the result that they are unable to recover all or some of their investment). These contexts are marked, moreover, by the sudden loss of meaning of both economic calculation (with the uncertainty this brings) and, as a result, previously existing social bonds. Mimicry intensifies, generating greater violence. In this light, experiences of hyperinflation and bankruptcy perfectly illustrate how, behind these turbulent macroeconomic movements, we find concrete social configurations, and intense mimetic reflexes in the behavior of individuals. And as these authors go on to argue, in such moments of monetary chaos and serious social conflict, institutions are required to take crucial decisions, the State adopting one or other type of measure in function of the balance of power existing in a given society. Thus, the State may establish a hierarchy of right to recover debts, set debt reductions, limits or, as is often the case at present, socialize the losses. These attempts to resolve economic and financial crises generate a series of social dynamics in which violence, whether real or symbolic, explodes in one way or another. In extreme circumstances, this can cause the implosion, if not of the capitalist system itself, at least of the financial system, as happened in 1929 and, albeit to a more limited extent, in September 2008. However, in the case of this latest financial crisis and, indeed, its predecessors, the violence generated has been channeled in a very specific direction, far from the market and right at the heart of society.

3. The Resolution of the Crisis as Sacrifice

One financial crisis has followed another over time, but the financial world is still not just alive, but in fact continues to thrive, nourished by the many corpses that it leaves in its wake. What is its magic formula for survival? Why is it that, despite the financial sphere being responsible for these crises, the violence they generate does not sweep it away? In this respect, again, René Girard's work provides a timely theoretical framework to help us to reconstruct a way out, if only partial, of the crisis. The key is his theory of the scapegoat (see 1986; for further analysis on Girard's theory, see Fleming 2004). According to Girard (1986), the principle of rivalry dominates in all spheres of experience, and when unleashed, it sows confusion and self-damage to social groups. This might lead to crises that represent a cathartic solution to the aforementioned rivalry. Over the course of his complex and dense works, Girard (1977, 1986) argues that crises in society are also marked by strong doses of violence; they imply the collapse of the existing institutions, which, in other eras, could mean the peace of the clan or of the medieval cities hit by a disaster or a calamity such as poor harvests or pandemics. Girard (1986) shows how, in some of the classic examples, outbreaks of diseases such as the plague led to an increase in violence as a result of the inability of the existing institutions to provide a response. The solution to the crisis was, and, in his opinion, usually has been, the same. First, some possible culprits, even if innocent, are identified (in the case cited by Girard, the victims are the Jews). This paves the way for their persecution by the others, which ends with the genocide or expulsion of the chosen victims, the scapegoats (in this case, the murdered Jews). Violence provides a liberating catharsis, which makes it possible to restore the damaged bonds within the community

[4] In this way, when savers withdraw their money *en masse* from a doubtful financial institution, the ultimate effect is obviously the suspension of payment. An example in relation to this type of collapse of savers' confidence is the excellent analysis of the nationalized British building societies by Klimecki and Willmott (2009).

by collectively sacrificing the victim, the violence then abating until the next crisis. The scapegoat is, therefore, the innocent victim who becomes the focus of the hatred of all (Girard 1987). The function of this collective crime is to reconstruct social relations, to the point that Girard (1977) even states that, without the sacrificial crisis, and given that violence can always be stirred by mimetic rivalry, the community would destroy itself completely. The violence of the sacrifice provides the basis for reestablishing order and the appearance of a sphere of the sacred, paradoxically in the form of religious adoration of a totem, god or symbol derived from the scapegoat mentioned above (Girard 1986, 2001, 2003).

Curiously, Girard's approach provides a rich and suggestive framework for our analysis of financial crises. The scapegoat is essential to any understanding of how neoliberal policies operate. As noted above, this fits in well with an image of the financial world driven by intense mimetic relations among the agents who operate within it: they all seek to get richer just like the rest, seeing others (banks, brokers) as rivals and models to follow. In this ruthless competition, agents assume ever more antagonistic positions, tied together in relations, which, in the case that concerns us here, take the form of relations between creditor and debtor which have permeated the whole socioeconomic world of new, debt-driven capitalism. Mimicry holds sway among the financial agents who, in order to outdo their rivals, also take increasingly risky financial decisions (whether entrusting their money to less solvent agents or purchasing more expensive houses). This gives rise to a curious paradox whereby the greater the agreement of the agents with respect to the supreme value (in this case, the idol of money), the greater the risk of self-destruction as a result of a mimetic paroxysm in which all rationality with respect to economic decision-making is abandoned. Eventually, however, the limits of this commotion are exposed when the crisis hits. And here, the figure of the scapegoat appears to offer an extremely useful concept through which to make sense of both the current economic crisis and the violence it embodies.[5]

4. Sacrifice at Play: The Debt Crisis in Southern Europe

Many scholars are currently debating over the origins and outcomes of the latest financial crisis from a socioeconomic perspective (e.g., Lounsbury and Hirsch 2010; McKenzie 2011; Castells et al. 2012; Van Der Zwan 2014; Dinerstein et al. 2014; Coleman and Tutton 2017). The Great Recession in Europe (but, before that, in Latin America, and many other places) has revealed the true scale of this symbolic violence with real economic effects (see, e.g., Lapavitsas et al. 2012), with collapses of grotesque real estate bubbles, evictions, massive bank frauds, mass unemployment, cuts in social and labor rights, falling wage levels, indiscriminate increases in prices and taxes and police violence against peaceful demonstrators. The dramatic increase in social inequalities simply confirms that the current crisis has not affected all social groups equally: the senior executives of the major global financial institutions, recipients of multimillion dollar salaries and representatives of a privileged lifestyle in obscene contrast to the condition of most of the planet's inhabitants, have escaped virtually scot-free except for criticism of their bonuses, despite the strong implications of their limited liability. Some scholars have argued that this represents a moral hazard that could explain, at least partly, the systemic instability of contemporary capitalism (see Djelic and Bothello 2013). It is not just that their businesses have been rescued by the taxpayer, but that they have embarked on a brutal defense of their interests which, in many cases, and above all in Europe, has led to the implementation of drastic neoliberal ideas and policies (Lapavitsas 2009). This defense of the interests of financial capital has acquired unprecedented force, venturing deep into previously unexplored terrain (in terms of cuts and privatizations) in order to further enrich the rich, even if this necessarily makes the poor even poorer (Chang 2010). In the face of the entrenchment of the financial caste, acting with the undeniable complicity of the political class in their

5 For reasons of space and relevance, here, we will focus on analyzing the scapegoat in the current economic crisis. Nonetheless, a quick glance at the past is sufficient to confirm that, in the case of conservative, reactionary or plainly fascist responses to the economic crisis, we always find scapegoats of one kind or other: Jews, subversives, immigrants, etc.

brutal reinforcement of the status quo, more and more social groups and sectors are experiencing greater vulnerability due to the loss of rights, falling living standards or diminishing incomes. And all these sacrifices are made to placate a series of entities, such as the financial markets and the credit-rating agencies, whose negative reactions in the form of higher-risk premiums or lower credit ratings have the potential to provoke collective bankruptcy.

What brought about the 2008 financial crisis? The fundamental cause was the inability of some financial agents to meet their obligations to their creditors after entering into exceedingly complex and high-risk financial operations induced by the mimetic effect. For instance, the behavior of the Spanish financial institutions during this latest crisis seems to fit this pattern (see López and Rodríguez 2011; Buendía 2020). They try to compete with their rivals in the national and European markets at the same time as they mimic their strategies, with all these manifested as high exposure in the real estate market, massive debt to foreign financial institutions, astronomical financial rewards for senior executives, the same declarations regarding the solvency of the Spanish financial system and so forth. The 2008 crisis gave rise to situations charged with great symbolic violence (a possible collapse of the banking system with possible loss of almost everybody´s savings and investments), forcing the mobilization of vast public resources that consecrated the doctrine of "too big to fail" and the definitive hegemony of the financial sector in the crisis. The citizens were indignant and directed their anger towards the bankers/brokers, even if the majority were participating, through multiple networks of debt, in the system and, for that reason, had an interest in its survival. In these circumstances, the financial world and its champions seek out different scapegoats at different levels. In the United States, the first scapegoats were the recipients of sub-prime mortgages, subsequently evicted; later, after the bailout of the financial institutions, it was the turn of different areas of public expenditure. In the European case, the alliance between the banks and some of the wealthiest citizens has managed to focus attention on other scapegoats, particularly the irresponsible Southern European States. Headed by Greece, the so-called PIGS (Portugal, Italy, Greece, Spain) stand accused of wasting their resources, not introducing the necessary structural reforms, failing to modernize despite receiving European Union funds, of having lived beyond their means, and so on. They have been forced, and their governments have agreed, to recognize their guilt, their enthusiastic adhesion to that decision which has annulled them in a script which appears taken literally from Girard (1987). For instance, the Spaniards (and, above all, the popular classes in Spain) have also been required to acknowledge that they have effectively lived above their means (Alonso et al. 2015), a highly popular discourse in other parts of Europe where austerity has reigned (see, e.g., Basu 2019). At the same time, the dominant classes have been able to retain their newly reinforced privileges, to take advantage of the revolving door between public and private sectors, to elude all types of ethical obstacles to the pursuit of personal profit and, in short, to accumulate more wealth in an unhindered way by dispossessing the rest of the population.

Within a space such as Southern Europe, the management of the crisis and the need to calm the calamities spat out by the sacred monster labeled *The Markets*, which, like some kind of 21st-century Tezcatlipoca, must be placated through constant sacrifices, have pointed at various scapegoats. The crisis has already seen a number paraded before the sacrificial pyre: the salaries and contracts of workers in the public and private sector; historic employment and union rights (the ferocity of the onslaught on the main union organizations, their leaders and shop stewards is particularly remarkable), health and education, but also citizens of flesh and blood who have lost everything. When the crisis struck, debts (private, then public) went unpaid and intense symbolic violence broke out; some (promoters, large construction companies) have been given financial support, while others, small, poor and indebted homeowners (scapegoats), have been forcibly evicted and dispossessed. This was not enough, however. The banks eventually had to be bailed out, and the search for a scapegoat has only intensified. The markets were to be placated before allowing any financing of Southern European countries' public debt, in order to keep the financial system afloat. Sacrifices were then offered to exit the crisis, and scapegoats were found. These were the citizens and particularly the popular classes, and symbolic violence was put into force, dismissing them from their jobs, depriving them of medical care, charging for basic services, cutting salaries and promotion prospects and making them accept the loss of their savings. All these steps were taken to persuade the markets to declare a truce, which, in the long run, serves to reconstitute the old sacred order of the financial market of interlinked debts a priori believed recoverable without undue difficulty.

In this way, the resolution of the crisis can take two forms. The first is the payment of what is owed, in many cases as a result of more or less irresponsible loans, but such obligation to the payment demonstrates the crucial role that debt still plays in the state of domination that is capitalism: States, companies (small- and medium-sized firms only, of course) and indebted households must pay off their debts, and remain chained to the system by this never-ending debt cycle. This implies becoming more competitive and working hard, innovating more, spending more. The second is through the necessary sacrifice of scapegoats which, as Girard (1986) showed, belong to obvious victim groups—in this case, the least privileged classes and public sector workers. The need to make sacrifice is defended by means of a political discourse that tells us that we have lived beyond our means, at the same time as the steady dribble of nationalizations in the financial sector continues. This coincides almost point by point with an ironic observation of Aglietta and Orléan (1982) nearly four decades ago: the recession is a cure applied after periods of excessive prosperity, a call for attention and for moderation directed at those who have been reckless enough to live above their possibilities—that is, at the workers and poor.

5. Discussion: The Crisis as Disciplinary Device

The crisis is a phenomenon whose impact always goes far beyond the strictly economic terrain: it is the detonator for the articulation of a new governmentality in the sense of a legitimized and embedded form of institutional domination of the population. As such, it entails the disciplinary adjustment of bodies to the production of mercantile value and the production of symbolic meaning of an ideological order which is restructured in each crisis. The interlinked notion of calculation and domination revealed by the crisis is effectively expressed in the concept of *dispositif* developed by Foucault (2002) and his followers. A *dispositif* is a normative network or framework which takes over human life, determining its forms of existence and shaping its behavior. Thus, paradoxically, while conventionally considered to be a disorder or maladjustment, in that the very use of the discourse tends to invoke the idea of struggle, or a decisive moment or key period in the evolution of an illness, the crisis in fact operates as an organizing idea (see Deleuze 1992). The crisis of the Great Recession is a good example of the increase in liberal biopower, because, as well as codifying an entire new subjectivity around mercantile individualization (expressed in every possible form of technological and cognitive renovation of the legitimacy of property and calculation), it represents the reinforcement of all the financial powers over and above any public, social, communitarian or cultural value. The neoliberal cycle has constituted a new governmentalization of the most genuine principles of capitalist reason which embeds the basic norms of the evaluation of private capital into the very existence and life projects of individuals (Foucault 2008; Lazzarato 2015). The logic of financial calculus spreads into an increasing range of social, economic and political policy domains (Bryan and Rafferty 2014; Komporozos-Athanasiou and Fotaki 2020). Power and control over the very lives of people become the strategic variable for the analysis of the crisis as a form of managing social conflict.

The sacrificial dimension is, in this context, an essential element of the typical and highly topical narrative of capitalist crises. It reveals how, in the pathological and therapeutic approach to the crisis propagated by the established powers, the threatened collapse of the system is always presented as the responsibility of a combination of external enemies who infect us (such as international trade, other economies, uncontrolled increases in the cost of raw materials or distant financial maneuvers) and enemies within who weaken and ruin us (the people who do not work hard enough, those of us who have lived above our means, the troublemakers, the spendthrifts, and so on). Politicians and media will actually mobilize nostalgic and socially conservative ideas of work and community in order to justify cuts to public services and create divisions between the deserving and undeserving (Forkert 2017). This surely helps to set up fears and expectations that constrain the public debate about the financial crisis (Peckham 2013). However, the one thing that does not change is the end of the story: demands for greater power and autonomy for the elites to facilitate the emergence of *an iron surgeon*, to make adjustments, "cure", amputate and revitalize the economic body. Capitalist crises in general, and this latest financial crisis in the most extreme way, have used the themes and rhetorical devices of horror stories, such as the inevitability of evil, fueling of unease, the crisis as a cruel and insatiable monster, to instill the greatest possible anxiety, very much in the way Foucault conceptualized "technologies of the self" (Foucault 1988). The same elements also serve to identify the necessary culprits—the Jews in many other historical crises, excessive public spending in this one—and set up the scapegoats, which here are represented as the welfare state or the poor mortgage and debt holders. This has helped to justify ever tougher measures, always in line with the code of values of the dominant power, and the regressive and infantilized acquiescence of the weakest social groups. These groups appear to have little choice but to accept the loss of rights, salaries, services and income, all in a desperate attempt to satisfy this rampaging financial *Moloch*[6]. Evidently, fear is the message (Altheide 2002), amplified ad infinitum by the media (ever more deeply embedded in the logic of capitalism itself); and, for that very reason, the solution is punishment for those who try to live outside the code of economic power (that is, outside the logic of the market). As we have seen very clearly in the Great Recession, all those impure elements which have proliferated in the social or public phase of the economic cycle can be sacrificed to the market, the great, recurrent totem of capitalist modernity. This cycle will serve as a public warning and example to all those who are presented as if they were free from the control and discipline of mercantile reason (civil servants, dependent population, benefit claimants, workers in non-profit cultural activities, and so on).

Thus, the crisis has been permanently present in the modern project, precisely because of the ambivalent and contradictory nature of this project, always split between its dimension of social progress and civilizing development and the perpetual return to the centrality of the accumulation of capital by all means possible (Bauman 2005). For that reason, the crisis tends to become socially global, in that, whilst it usually breaks out in the subsystem of the economy and accumulation, it rapidly spreads and destabilizes the other subsystems (politics, legitimacy, culture, values, etc.) because the economy provides with the central meaning for the modern narrative and its reason. If capitalism is, as Joseph Schumpeter (2010) claimed, a process of permanent creative destruction, the crisis represents the permanent restructuring and the institutionalized and manipulated uncertainty which forms part of the life story of social groups. The crisis, therefore, is how societies experience life according to a code of values which develops from the way social conflict is approached and managed. Just as progress and the spirit of conquest characterized the dominant discourses of classical industrial modernity, uncertainty, risk and fear create the form of social construction of the experience of time in late and reflexive modernity (Lash and Urry 1994; Revault D'Allonnes 2012). Therefore, like any process of framing the social construct, the crisis as discourse models subjectivities in accordance to the dictates of the dominant disciplinary order, using the concept of discipline as a way of adjusting bodies to the production of mercantile reason (Lazzarato 2015).

Thus, the crisis has operated as an increasingly complex disciplinary *device* dressed up in different discursive forms, whether natural, medical, biological or prophylactic, from the very origins of modernity to the current technological, cybernetic, economic and financial forms, but always resulting in the use of bodies controlled and regulated by the twin

6 This logic is behind other policies deployed by the neoliberal governmentality—for example, the punishment of the poor (Wacquant 2009).

concepts of utility/docility. This obviously echoes Foucault's ideas (Foucault 1991) on discipline and punishment in modernity. The omnipresence of the notion of crisis and its obligations (effort, sacrifices to get over it, hard and heartless curative measures) in modernity indicate that political economy has taken control of disciplinary technologies and become the matrix of all actions, justifications and required or rather demanded forms of behavior. The discourse (or resource) of the crisis can thus be seen to be one of the most powerful technologies of the self, which turns the prescriptions of the governmentality of power into a subjectively perceived necessity and a reasoned, reasonable, and even voluntary, individual response.

6. Conclusions

The main contribution of this paper is to highlight the relevance of sacrifice in modernity, by focusing on its role in the economic crises, some of the most relevant events in modern times, as their outcomes will affect substantially the fate of many. Drawing on a number of theories, particularly the concept of sacrifice proposed by René Girard, we aimed at explaining the sacrificial logics behind the deployment of the crisis and illustrated its pertinence by looking at the well-known effects of the Great Recession in the context of Southern Europe. The importance of the role of debt in these logics of sacrifice has also been highlighted, once debt usually triggers financial crises. The metaphor of the scapegoats is revealing, once we have just witnessed how the popular classes of many countries have paid a disproportionate price for the wrongdoing of the financial sector, in the form of unemployment, cuts in welfare and medical services and loss of rights.

However, there are limitations in this approach. Girard's theory of sacrifice is controversial and has been criticized by scholars such as Hénaff (2012), who considers his definition of sacrifice ill-based as it does not take into account the *offering* involved in any act of sacrifice. The economic crises can be the result of various different factors—for instance, a phenomenon outside the market exchanges or the financial sector: the coronavirus crisis is an excellent example, as a pandemic can trigger an unexpected downturn for the global economy. The outcomes of the crises may also lead to diverging scenarios, including changes that may actually benefit the poorer groups in society, as happened after the Great Depression of 1929. After all, the resolution of the crises may also depend on the balance of power in society. Despite these possible scenarios, in the neoliberal era, finances have greater power than ever, so it is likely that the logic of sacrifice might be deployed again in the terms we have described in our article. Given the current context, we will probably know it rather sooner than later.

Author Contributions: Both authors have contributed equally to the manuscript. Both authors have read and agreed to the published version of the manuscript.

Funding: This research was funded by the Spanish Ministry of Science and Innovation, grant number PGC2018-097200-B-I00.

Acknowledgments: We would like to thank Justin Byrne for the translation of the first versions of this paper; Riie Heikkilä for her invaluable comments on the final manuscript; and the reviewers for their comments which helped to improve the paper.

Conflicts of Interest: The authors declare no conflict of interest. The funders had no role in the design of the study; in the collection, analyses, or interpretation of data; in the writing of the manuscript, or in the decision to publish the results.

References

Abend, Gabriel. 2014. *The Moral Background: An Inquiry into the History of Business Ethics*. Princeton: Princeton University Press.
Adey, Peter, Anderson Ben, and Graham Stephen. 2015. Introduction: Governing Emergencies: Beyond Exceptionality. *Theory, Culture & Society* 32: 3–17. [CrossRef]
Adkins, Lisa. 2017. Speculative futures in the time of debt. *The Sociological Review* 65: 448–62. [CrossRef]
Agamben, Giorgio. 2005. *State of Exception*. Chicago: The University of Chicago Press.
Aglietta, Michel, and André Orléan. 1982. *La violence de la monnaie*. Paris: Presses Universitaires de France.

Akerlof, George, and Robert J. Shiller. 2009. *How Human Psychology Drives the Economy, and Why It Matters for Global Capitalism*. Princeton: Princeton University Press.

Alonso, Luis Enrique, and Miguel Martínez Lucio. 2006. *Employment Relations in a Changing Society: Assessing the Post-Fordist Paradigm*. Bakingstoke: Palgrave Macmillan.

Alonso, Luis Enrique, Fernández Rodríguez, Carlos J, and Rafael Ibáñez Rojo. 2015. From consumerism to guilt: Economic crisis and discourses about consumption in Spain. *Journal of Consumer Culture* 15: 66–85. [CrossRef]

Altheide, David L. 2002. *Creating Fear: News and the Construction of a Crisis*. Hawthorne: De Gruyter.

Aristotle. 1999. *Politics*. Kitchener: Batoche Books.

Basu, L. 2019. Living within our means: The UK news construction of the austerity frame over time. *Journalism* 20: 313–30. [CrossRef]

Bauman, Zygmunt. 2005. *Modernity and Ambivalence*. Ithaca: Cornell University Press.

Berger, Peter. 1974. *Pyramids of Sacrifice: Political Ethics and Social Change*. Harmondsworth: Penguin Books.

Beriain, Josetxo. 2017. Las metamorfosis del don: Ofrenda, sacrificio, gracia, substitute técnico de Dios y vida regalada. *Política y Sociedad* 54: 645–67. [CrossRef]

Bourdieu, Pierre. 1991. *Language and Symbolic Power*. Cambridge: Polity Press.

Bourdieu, Pierre. 2000. *Pascalian Meditations*. Cambridge: Polity Press.

Bryan, Dick, and Michael Rafferty. 2014. Financial Derivatives as Social Policy beyond Crisis. *Sociology* 48: 887–903. [CrossRef]

Buendía, Luis. 2020. A perfect storm in a sunny economy: A political economy approach to the crisis in Spain. *Socio-Economic Review* 18: 419–38. [CrossRef]

Castells, Manuel, Caraça João, and Cardoso Gustavo, eds. 2012. *Aftermath: The Cultures of the Economic Crisis*. Oxford: Oxford University Press.

Chang, Ha-Joon. 2010. *23 Things They Don't Tell You About Capitalism*. London: Penguin.

Coleman, Rebecca, and Richard Tutton. 2017. Introduction to special issue of Sociological Review on 'Futures in question: Theories, methods, practices'. *The Sociological Review* 65: 440–47. [CrossRef]

Daguerre, Anne. 2014. New corporate elites and the erosion of the Keynesian social compact. *Work, Employment and Society* 28: 323–34. [CrossRef]

Deleuze, Gilles. 1992. What Is a Dispositif? In *Michel Foucault Philosopher*. Edited by Timothy J. Armstrong. Hemel Hempstead: Harvester Wheatsheaf, pp. 159–68.

Dinerstein, Ana C., Schwartz Graham, and Taylor Gregory. 2014. Sociological imagination as social critique: Interrogating the 'global economic crisis'. *Sociology* 48: 859–68. [CrossRef]

Djelic, Marie-Laure, and Joel Bothello. 2013. Limited liability and its moral hazard implications: The systemic inscription of instability in contemporary capitalism. *Theory and Society* 42: 589–615. [CrossRef]

Durkheim, Émile. 1960. *Les Formes élémentaires de la vie religieuse*. Paris: PUF.

Eberhart, Christian, and Donald Schweitzer. 2019. The Unique Sacrifice of Christ According to Hebrews 9: A Study in Theological Creativity. *Religions* 10: 47. [CrossRef]

Fama, Eugene F. 1970. Efficient capital markets: A review of theory and empirical work. *Journal of Finance* 25: 383–417. [CrossRef]

Fleming, Chris. 2004. *René Girard: Violence and Mimesis*. Cambridge: Polity.

Forkert, Kirsten. 2017. *Austerity as Public Mood: Social Anxieties and Social Struggles*. London: Rowman & Littlefield International.

Foucault, Michel. 1980. *Power/Knowledge: Selected Interviews and Other Writings, 1972–1977*. New York: Pantheon Books.

Foucault, Michel. 1988. *Technologies of the Self: A Seminar with Michel Foucault*. Amherst: University of Massachusetts Press.

Foucault, Michel. 1991. *Discipline and Punish: The Birth of the Prison*. London: Penguin.

Foucault, Michel. 2002. *The Archaeology of Knowledge*. London: Routledge.

Foucault, Michel. 2008. *The Birth of Biopolitics: Lectures at the Collège de France, 1978–1979*. New York: Palgrave MacMillan.

Galbraith, John Kenneth. 1994. *A Short Story of Financial Euphoria*. New York: Penguin.

Girard, René. 1977. *Violence and the Sacred*. Baltimore: The Johns Hopkins University Press.

Girard, René. 1986. *The Scapegoat*. Baltimore: The Johns Hopkins University Press.

Girard, René. 1987. *Job: The Victim of His People*. Stanford: Stanford University Press.

Girard, René. 2001. *I See Satan Fall Like Lightning*. Maryknoll: Orbis Books.

Girard, René. 2003. *Le Sacrifice*. Paris: Bibliothèque nationale de France.

Graeber, David. 2011. *Debt: The First 5000 Years*. New York: Melville House Publishing.

Hackworth, Jason. 2012. *Faith Based: Religious Neoliberalism and the Politics of Welfare in the United States*. Athens: The University of Georgia Press.

Harvey, David. 2005. *A Brief History of Neoliberalism*. Oxford: Oxford University Press.

Harvey, David. 2010. *The Enigma of Capital: And the Crises of Capitalism*. Oxford: Oxford University Press.

Hénaff, Marcel. 2002. *The Price of Truth: Gift, Money, and Philosophy*. Stanford: Stanford University Press.

Hénaff, Marcel. 2012. Three Crucial Aspects of Religion in Human Evolution: Shamanism, Sacrifice and Exogamic Alliance. *European Journal of Sociology* 53: 327–35. [CrossRef]

Hubert, Henri, and Marcel Mauss. 1981. *Sacrifice: Its Nature and Functions*. Chicago: Chicago University Press.

Kahneman, Daniel. 2003. A Perspective in Judgement and Choice: Mapping Bounded Rationality. *American Psychologist* 58: 697–720. [CrossRef]

Kindleberger, Charles P., and Robert Z. Aliber. 2005. *Manias, Panics and Crashes: A History of Financial Crises*. New York: Palgrave MacMillan.
Klimecki, Robin, and Hugh Willmott. 2009. From Mutualization to Meltdown: A Tale of Two Wannabe Banks. *Critical Perspectives on International Business* 5: 120–40. [CrossRef]
Koch, Max. 2011. *Capitalism and Climate Change. Theoretical Discussion, Historical Development and Policy Responses*. Basingstoke: Palgrave MacMillan.
Komporozos-Athanasiou, Aris, and Mariana Fotaki. 2020. The imaginary constitution of financial crises. *The Sociological Review* 68: 932–47. [CrossRef]
Krippner, Greta R. 2011. *Capitalizing on Crisis: The Political Origins of the Rise of Finance*. Cambridge: Harvard University Press.
Lapavitsas, Costas. 2009. Financialised Capitalism: Crisis and Financial Expropriation. *Historical Materialism* 17: 114–48. [CrossRef]
Lapavitsas, Costas, Kaltenbrunner Annina, Labrinidis George, Lindo Duncan, Meadway James, Mitchell Jo, Painceira Juan Pablo, Powell Jeff, Pires Eugenia, Stenfors Alexis, and et al. 2012. *Crisis in the Eurozone*. London: Verso.
Lash, Scott, and John Urry. 1994. *Economies of Signs and Space*. London: TCS/Sage.
Lazzarato, Maurizio. 2012. *The Making of the Indebted Man: An Essay on the Neoliberal Condition*. Los Angeles: Semiotext(e) Intervention Series.
Lazzarato, Maurizio. 2015. Neoliberalism, the Financial Crisis and the End of the Liberal State. *Theory, Culture & Society* 32: 67–83. [CrossRef]
López, Isidro, and Emmanuel Rodríguez. 2011. The Spanish Model. *New Left Review* 69: 5–29.
Lordon, Frédéric. 2010. *Capitalisme, désir et servitude. Marx et Spinoza*. Paris: La Fabrique éditions.
Lounsbury, Michael, and Paul Hirsch, eds. 2010. *Markets on Trial: The Economic Sociology of the U.S. Financial Crisis*. Bingley: Emerald.
Mauss, Marcel. 1979. *Sociología y antropología*. Madrid: Tecnos.
McKenzie, Donald. 2011. The Credit Crisis as a Problem in the Sociology of Knowledge. *American Journal of Sociology* 116: 1778–841. [CrossRef]
Montier, James. 2000. *Behavioral Finance: Insights into Irrational Minds and Markets*. Chichester: Wiley.
North, Douglass C., Wallis John Joseph, and Weingast Barry R. 2009. *Violence and Social Orders: A Conceptual Framework for Recorded Human History*. Cambridge: Cambridge University Press.
Peck, Jamie. 2013. Austere reason, and the eschatology of neoliberalism's End Times. *Comparative European Politics* 11: 713–21. [CrossRef]
Peckham, Robert. 2013. Economies of contagion: Financial crisis and pandemic. *Economy and Society* 42: 226–48. [CrossRef]
Poppe, Christian, Lavik Randi, and Elling Borgeraas. 2016. The dangers of borrowing in the age of financialization. *Acta Sociologica* 59: 19–33. [CrossRef]
Reinhart, Carmen M., and Kenneth S. Rogoff. 2009. *This Time Is Different: Eight Centuries of Financial Folly*. Princeton: Princeton University Press.
Revault D'Allonnes, Myriam. 2012. *La crise sans fin: Essai sur l'expérience moderne du temps*. Paris: Seuil.
Schumpeter, Joseph A. 2010. *Capitalism, Socialism and Democracy*. London: Routledge.
Simmel, Georg. 1978. *The Philosophy of Money*. London: Routledge.
Sombart, Werner. 2001. *The Jews and Modern Capitalism*. Kitchener: Batoche Books.
Stockhammer, E. 2009. The finance-dominated accumulation regime, income distribution and the present crisis. *Papeles de Europa* 19: 58–81
Van Der Zwan, Natascha. 2014. Making Sense of Financialization. *Socio-Economic Review* 12: 99–129. [CrossRef]
Wacquant, Loïc. 2009. *Punishing the Poor: The Neoliberal Government of Social Insecurity*. Durham: Duke University Press.
Weber, Max. 2005. *The Protestant Ethic and the Spirit of Capitalism*. London: Routledge.

MDPI

Article

Trauma and Sacrifice in Divided Communities: The Sacralisation of the Victims of Terrorism in Spain

Eliana Alemán [1,*] and José M. Pérez-Agote [2,*]

[1] Departament of Sociology and Social Work, Universidad Pública de Navarra, 31006 Pamplona, Navarra, Spain

[2] I-Communitas, Institute for Advanced Social Research, Universidad Pública de Navarra, 31006 Pamplona, Navarra, Spain

* Correspondence: eliana.aleman@unavarra.es (E.A.); jose.perez.agote@unavarra.es (J.M.P.-A.)

Abstract: This work aims to show that the sacrificial status of the victims of acts of terrorism, such as the 2004 Madrid train bombings ("11-M") and ETA (Basque Homeland and Liberty) attacks in Spain, is determined by how it is interpreted by the communities affected and the manner in which it is ritually elaborated a posteriori by society and institutionalised by the state. We also explore the way in which the sacralisation of the victim is used in socially and politically divided societies to establish the limits of the pure and the impure in defining the "Us", which is a subject of dispute. To demonstrate this, we first describe two traumatic events of particular social and political significance (the case of Miguel Ángel Blanco and the 2004 Madrid train bombings). Secondly, we analyse different manifestations of the institutional discourse regarding victims in Spain, examining their representation in legislation, in public demonstrations by associations of victims of terrorism and in commemorative "performances" staged in Spain. We conclude that in societies such as Spain's, where there exists a polarisation of the definition of the "Us", the success of cultural and institutional performances oriented towards reparation of the terrorist trauma is precarious. Consequently, the validity of the post-sacrificial narrative centring on the sacred value of human life is ephemeral and thus fails to displace sacrificial narratives in which particularist definitions of the sacred Us predominate.

Keywords: sacrifice; cultural trauma; victims of terrorism; ritual; performance

Citation: Alemán, Eliana, and José M. Pérez-Agote. 2021. Trauma and Sacrifice in Divided Communities: The Sacralisation of the Victims of Terrorism in Spain. *Religions* 12: 104. https://doi.org/10.3390/rel12020104

Academic Editors: Javier Gil-Gimeno, Josetxo Beriain and Celso Sánchez

Received: 15 January 2021

Accepted: 30 January 2021

Published: 4 February 2021

Publisher's Note: MDPI stays neutral with regard to jurisdictional claims in published maps and institutional affiliations.

1. Introduction: Sacrifice, Terrorism and Ritual

In this article, we seek to show that the sacrificial status of the victims of acts of terrorism such as the 2004 Madrid train bombings (referred to in Spanish and hereinafter as 11-M) and ETA (Basque Homeland and Liberty) attacks in Spain does not occur in a closed form but is subject to different interpretations by those affected by the attacks. We argue that the sacrificial sacralisation of the victims is, in reality, ritually elaborated a posteriori by society and institutionalised by the state. We also explore the way in which the sacralisation of the victim is used in socially and politically divided societies to establish the limits of the pure and the impure in the dispute regarding the definition of the "Us".

To demonstrate this thesis, we hermeneutically address the relationship between sacrifice and terrorism, highlighting the process whereby social representations of the victims are elaborated. In doing so, we shall make reference to two types of empirical reality drawn from Spanish society. On the one hand, we describe two traumatic events of particular social and political significance (the case of Miguel Ángel Blanco and 11-M), which are the subject of collective processes of a cultural nature. On the other hand, we analyse different manifestations of the institutional discourse on victims in Spain, examining their representation in legislation, in public demonstrations by associations of victims of terrorism and in commemorative "performances" staged in Spain.

Ever since Hubert and Mauss published their seminal study in 1899, sacrifice has come to be seen as a ritual act in which the participants take on a sacred nature, in such a way that the sacrificial offering passes from the commonplace to the religious. This is a central element of the religious experience and is characteristic of Neolithic agricultural cultures and societies. Although it entails a complex procedure that can take multiple forms and be used for the most diverse purposes, it invariably establishes "a means of communication between the sacred and the profane worlds through the mediation of a victim, that is, of a thing that in the course of the ceremony is destroyed" (Hubert and Mauss 1899, p. 76). Thus, the total or partial destruction of the offering, albeit only in vegetable form, is an essential feature of the sacrifice.

However, in order for the victim to pass from the common to the religious domain, it must be consecrated by means of a series of ritual operations that purify it, so that it will act as an intermediary between the *sacrifier*[1]—the individual or collective subject to whom the benefits of sacrifice accrue—and the divinity to whom the sacrifice is generally made. The success of the sacrificial ritual, argue Hubert and Mauss (1899), depends on each of the steps assigned to the participating elements—*sacrifier*, sacrificer, instruments, victim and divinity—performing their role in perfect continuity, without interruption and in the prescribed order. Otherwise, the powerful and destructive forces involved would turn against both the sacrifier and the sacrificer.

Evidently, the sacrificial narrative that Hubert and Mauss describe cannot be precisely mapped to the present day. Social and civilisational changes have transformed the role of ritual in general and sacrifice in particular. The sacrificial logic they describe, centring as it does on the immolation of the victim, would later be rejected by Christianity. On the grounds that "it is criminal to kill the victim because he is a sacred being", Christianity counters with a post-sacrificial narrative centring on "the anti-sacrificial sacralisation of the human individual", from whose secularised re-interpretation we get the human rights legislation of modern societies, as described by Durkheim (1973) and Joas (2019), Beriáin (2017, pp. 658–59). At the same time, despite broad consensus throughout the twentieth century on Hubert and Mauss's work, a number of alternative explanations have been postulated, such as Girard's influential hypothesis of substitution. Girard extends the anti-sacrificial narrative to Neolithic societies, arguing that "There is no question of 'expiation.' Rather, society is seeking to deflect upon a relatively indifferent victim, a 'sacrificeable' victim, the violence that would otherwise be vented on its own members" (Girard 2005, p. 8).

In determining the role of sacrifice in modern societies, we need to consider two important issues. Firstly, in such societies, the ritual form loses its capacity to perform its social function effectively and is thus replaced by a cultural performance. According to Jeffrey Alexander, in less differentiated and complex societies, the participation of community members in the ritual regenerates social cohesion. However, in more complex and differentiated contemporary modern societies, the ritual is not capable of maintaining the fusion in itself; social performance is the type of experience that can re-fuse the social elements that have lost their cohesion (Alexander 2006, 2017). This, however, does not mean that the narrative of the sacrificial ritual vanishes when the practice of cultural performance is extended. On the one hand, this is no more than a quasi-ritual practice; and on the other, contemporary social agents keep alive the narrative of the sacred significance of the ritual in many of their social practices. In the political domain, moreover, social performance requires the establishment of binary differences between a pure sacred and civic "Us", and its opposite, a contaminated, profane and anticivic "Them", as well as the emergence of heroic figures capable of making audiences feel the authenticity of these binary differences until re-fusion is attained. Achieving that impression of authenticity is a requisite for convincing the audience.

1 TRANSLATOR'S NOTE: I have used the term "sacrifier", coined by the English translator of Hubert and Mauss's original work, as a translation for *sacrificante* in Spanish (*sacrifiant* in the original French), for which there is no exact English equivalent.

Secondly, modern societies have introduced new variants of the sacred, which coexist and compete with former representations of the divinity. Thus, at different moments in time, realities such as the nation, the proletariat, the individual and democracy have all been sacralised. Elsewhere in this collection, Josetxo Beriáin makes reference to this aspect when he describes the clash that occurs in these societies between the sacrificial and the anti-sacrificial narrative (Beriáin 2021). From a post-Durkheimian perspective, this is endorsed by Bernhard Giesen when he links the collective identity to "the sacred", denoted by extraordinary moments or experiences that escape ordinary classification because of their exceptional capacity for transcendence. "The sacred stands for the collective identity of a social community" when the way in which the self-referential narrative of the collective self interprets its collective identity does not square with the real and visible representations of this collective self (Giesen 2006, p. 329). Ritual and performance communicate the sacred world of the collective identity with the profane world of their real and visible representations, thus constructing that "elementary communitas that transcends social cleavages and unites the body social" (Giesen 2006, p. 342).[2]

In order to understand how society and the state sacralise the victims of terrorism, we need to consider the role of all the elements taking part from the perspective of ritual, paying particular attention to the victim and to the way in which the sacred—on whose behalf this offering is made—is defined. In today's societies, dynamically diverse definitions of the sacred exist alongside one another, and their hierarchical status varies depending on the intensity and authenticity of the ritual performances in which they are activated, and the relationships binding the communities to which both victim and sacrificer belong.

The Girardian hypothesis of the scapegoat is often used to define the victim. From our perspective, however, it is of only limited use, since when it appears, it does so in a non-sacrificial form, accompanying a socially more relevant interpretation, as Hénaff notes, arguing that not all victims are sacrificial[3]. Therefore, adds Hénaff, we must not confuse sacrifice with victimisation.[4] Although Christ was indeed a scapegoat, his crucifixion, presented by Girard as the sacrifice of an innocent victim drawing visibility for the first time to the arbitrariness and injustice of sacrifice (Beriáin 2017, p. 657), was in fact no more than the execution of an agitator at the hands of the political power, argues Hénaff. And he concludes: "The sacrifice of Christ only occurred in a posteriori readings in evangelical preaching" (Hénaff 2002, p. 425).

Here, we need to be specific on the notion of terrorism and its ritual dimension. Beyond the very generic mention of terrorism as political violence exercised in the name of a social or political cause, not only have experts in this phenomenon been incapable of reaching any consensus whatsoever as to its definition; the divergence has only widened in the last forty years (Schmid 2011; Zulaika and Douglass 1990, 1996, 2008). While recognising the difficulty of establishing a stable concept as a starting point, for practical purposes, this work accepts the distinction between "old" and "new" terrorism.[5] This does not, however, signify that we necessarily validate all the features commonly attributed to each type. We need only clarify that by "old terrorism", we mean the kind practised during the second half of the twentieth century, and by "new terrorism", we refer to the sort that emerged from the mid-1990s (with the attacks on the World Trade Center in New York in 1993 and

2 *Communitas*: a concept coined by Turner to designate a unique form of solidarity that occurs in the liminal phase of a transitional ritual between two positions of status. The liminal moment dissolves the normative constraints and levels status distinctions among ritual participants; it fosters the creation of the communitas, in which ritual participants are brought closer to the primordial and existential, distancing them from their cognitive ties to the structured, normative social order and fusing them in a simple community of equals (Alexander 2017, pp. 53–54; Turner 1969).

3 Nonetheless, in the idea of violence as an inherent attribute of sacrifice—implicit in the scapegoat—Giesen sees a resource for awarding the supreme authenticity required for success of the ritual (Giesen 2006, p. 337).

4 "Sacrifice is an act that presupposes an addresser and an addressee. It is, above all an irreversible offering to an invisible beneficiary whose response must be obtained, hence the immolation. As in ceremonial gift-giving, something of one's own must be offered—hence the choice of a domestic animal. The fact that the immolated victim is called a victim does not mean that every victim is a sacrificed being. Lynchings, executions, or massacres are not sacrifices but victimization procedures" (Hénaff 2002, pp. 424–25).

5 See Blanco and Cohen (2016) about the distinction between "old" and "new" terrorism.

on the Tokyo subway in 1995), which took on a whole new dimension following the attacks of 11 September 2001. The forms that fall under the category of "old terrorism" act in the name of different causes, including religion, class and ethnic or national community. Of all of these, ethno-terrorism is especially significant (Zulaika and Douglass 1996). In contrast, the category of new terrorism is largely monopolised by Islamic fundamentalism.

Insofar as terrorism is an extremely polyhedral phenomenon, it cannot be reduced to a single, univocal pattern of action. Nonetheless, it contains a markedly ritualistic dimension within which one can identify participants and rites of passage and exit that frame the attack as a hierophanic element in which the sacred is manifested (Eliade 1985). In our case, the sacred coincides with the collective identity, in the form of the nation (for ETA) and the ummah or community of believers (for Al-Qaeda). The attack is the symbolic event that breaks the banal continuity of social order and their everyday secular lives (Giesen 2006, p. 327). It is "a ritual action that introduces a discontinuity into the ordinary course of events" (Zulaika 1991, p. 221). The fact that the victim is selected at random guarantees his or her innocence, giving the event an inexplicable character for its audience; unable to find a causal rationality for the victim's immolation, the audience feels more vulnerable to terror (Zulaika 1991; Zulaika and Douglass 1996). For ETA's defenders, its "members are priestly figures charged with the sacred function of defending the homeland; for its detractors, in contrast, they are the compendium of the most irrational danger and most abhorrent bestiality" (Zulaika 1991, p. 224).

The fact that the diverse manifestations of terrorism have a ritual character does not mean per se that the terrorist act or attack constitutes a sacrificial ritual. For this to be the case, the immolated victim must represent the sacrifier, be it an individual or a community. The sacralisation of the sacrificial victim must symbolise an exchange in which something of value –the victim—taken from the secular domain is consecrated, enabling a communion between the sacrifier and the sacred, whose favour (i.e., the fusion between the sacrifier and the sacred divinity) it is hoped to receive in return. Hence the difficulty of establishing a correspondence between terrorism and sacrifice, a problem that not all approaches to this phenomenon have been able to resolve.[6]

Nevertheless, the sacrificial sacrality of the victim of a terrorist attack may be subject to different interpretations. When the prevailing feature in the community from which the victim is taken is any particularist variant of a sacred "Us"—such as the nation—the immolation is not sacrificial, since it is not an offering by the community to which he or she belongs. However, the need to repair the cultural trauma caused makes it necessary to sacrificially sacralise the victim, in such a way that this community can maintain a quasi-sacrificial narrative and transmute itself a posteriori into the sacrifier community, or to appeal to the post-sacrificial narrative by making the sacralisation of the universalised human individual prevail over any other form of particular sacralisation. For the terrorist sacrificer, on the other hand, the immolation of the victim is only sacrificial in two cases. The first is when the sacrificer himself faces the risk of losing life or liberty or when he self-immolates, finding death. In this case, he is sacralised by his own community, but the target victim of his attack is not. The second case is when the victim is assigned the role of the scapegoat of the sacrifier community. In all other cases (the majority), it is seen simply as a necessary process of victimisation to achieve a specific goal.

2. Traumatic Events in a Divided Society

We cannot understand the reaction of Spanish society to terrorist attacks without considering the social and political polarisation forged by an original cultural trauma[7] that has not found civil reparation and which has fed and been fed by the superimposition of

[6] See, for example, the interpretation of the Real IRA attacks in Omagh (Ireland) and of ETA's killing of a bus driver in Itziar (Basque Country) in Dingley and Kirk-Smith (2002).

[7] "Cultural trauma occurs when members of a collectivity feel they have been subjected to a horrendous event that leaves indelible marks upon their group consciousness, marking their memories forever and changing their future identity in fundamental and irrevocable ways" (Alexander 2004).

different forms of violence exercised against different types of victim.[8] The transition to democracy after the period of Civil War and subsequent dictatorship of the victor, General Franco, did not result in a reparation for the victims from the losing side, nor for their families, who are still battling to have their remains—which were dumped in mass graves—recovered and identified.[9] Indeed, during the transition period, a hegemonic memory was imposed on the Civil War, whereby all Spaniards were equally to blame for what happened, denying the traumatic nature of its origins (Izquierdo 2017). Nor have the victims of the repression meted out under the dictatorship as yet received any reparation.

2.1. The Kidnapping and Murder of Miguel Ángel Blanco by ETA

ETA (Basque Homeland and Liberty) was set up in 1959 as an armed national liberation movement (Pérez-Agote 1984).[10] It aspired to the regeneration of Basque nationalism under the Franco dictatorship and rejected the appeal to race and religion traditionally defended by the Basque Nationalist Party (PNV). Amongst its creeds, it included socialism and a defence of the working class (Gurruchaga 1988; Zulaika 1991). ETA carried out its first assassination in 1968. By July 1997, it had killed 776 victims (De la Calle and Ignacio 2004). During the closing years of the Franco dictatorship, its actions were viewed with sympathy by opponents of the regime throughout Spain. This was especially true of its assassination of Admiral Carrero Blanco, Franco's prime minister and most likely successor, in 1973.

ETA emerged as a utopian and transgressive force in opposition to the violence of the state, notes Tejerina (2015). For many years, it primarily targeted members of the security forces, thus garnering strong support from radical nationalist sectors in the Basque Country, well into the period of the democratic transition (1975–1982). At the same time, however, it entirely lost any backing it had had elsewhere in Spain, and with the conclusion of the transition, its support among Basque nationalist circles also began to wane.[11] The goal of its strategy, according to De la Calle and Ignacio (2004), was to force the state to accept its demands and to control the population; it tried to reinforce this control by spearheading mobilisation for all kinds of social causes (Tejerina 2015). In short, a division was already becoming apparent in Basque society between those who supported ETA violence, those who rejected it and those who remained ambivalent, who embodied certain contradictions between rational and sentimental assessments (Pérez-Agote 1984).

In 1995, ETA began a new strategy, called the "socialisation of suffering", whereby it widened its repertoire of potential victims to extend the fear to society as a whole. This ended up catalysing mobilisation against the group (Gorospe 2018). Until 1997, the growing social opposition, in which pacifist movements and victims' associations (the first was created in 1981) played an important role (Mateo Santamaría 2018), had only a limited impact. However, everything changed that summer, when ETA lost its social legitimacy, thus beginning the road to the cessation of its armed activities (2011) and its ultimate dissolution (2018), leaving a final toll of 853 deaths.

On Thursday 10 July 1997, ETA announced that it had kidnapped Miguel Ángel Blanco, a councillor for the People's Party (Partido Popular) in Ermua, a small town in the Basque province of Biscay. ETA handed the government—and the People's Party—an ultimatum, giving it 48 h to bring the 600 ETA prisoners dispersed in prisons throughout Spain back to the Basque Country.[12] The action came in direct response to an operation ten days before by the security forces, in which they had successfully freed prison officer

---

8 On trauma in culturally divided communities, see Tognato (2013).

9 Although there is no consensus on the number of victims of the Civil War, recent research estimates that counting the combatants, those killed in the reprisal attack, executions and the victims of Franco's repression after the war, the death toll comes to around 600,000 (e.g., Preston 2012).

10 See De Pablo (2018) for the controversy on the date of ETA's foundation.

11 Support from radical nationalists in the Basque Country was strengthened by the dirty war waged against ETA by sectors of the security forces remaining from the dictatorship (González 2012; Woodworth 2001).

12 This measure, which is still in place, was intended by the state to hinder ETA's control over its prisoners.

José Antonio Ortega Lara, who had been held hostage by the group for over a year.[13] ETA acted out a ritual performance whose script envisaged two possible outcomes to the drama: either the victim would be executed or (an unlikely option) the prisoners would be relocated. Either would convincingly demonstrate to the audience, Spanish society, ETA's capacity to bend the state to its will. In effect, the government did not give in to the blackmail and when the deadline ran out, Miguel Ángel Blanco was found with his hands bound and two shots to the head. He died several hours later in hospital.

What the script that ETA had drawn up did not envisage was that from the very moment in which its communiqué was made public, it would trigger an overwhelming counter-performance driven by the uninterrupted coverage of the events in the media, which fed and was fed by an unprecedented social response. The media reported live on the progress of the search, declarations by a great range of public figures (including King Juan Carlos and the Pope) and citizen acts of protest. To facilitate the search, authorisation was also given to publish personal and family pictures and information on the councillor, leading to a complete identification by the audience with the victim. In short, the media provided the script of a counter-performance which served to sustain a climate of maximum intensity (catexis) over a two-day period and which also raised hopes that the victim might be saved.

Within this counter-performative framework, two sets of highly dramatic mobilisations were staged, constituting what Beriáin calls an expiatory rite of national mourning "that turned the secular death of an innocent into a grand sacred symbol" (Beriáin 2016, p. 109). The vigil of lit candles, played out on the night before the assassination in the square in front of the town hall in Ermua, reached a performative climax among all those protagonised by local people from the town, regardless of their political persuasions. In a study based on interviews with members of Elkarri and Gesto por la Paz, Funes tells how the vigil revived memories of the night before the last executions of the Franco era in 1975, and how the identification of ETA with Franco, for its intolerance and authoritarianism, "allowed the emotion of those hours to be augmented by emotions from the past" (Funes 1998, p. 102). It was, she adds, a catharsis that condensed in that event a process of change that was already underway and which entailed a broad social consensus with regard to the delegitimisation of violence for political ends (Sabucedo et al. 2000). The "spirit of Ermua" was born, a representation of the indignant and unanimous rejection of ETA violence by Spanish society and, in Basque society, the breaking of a spiral of silence. Throughout the rest of the country, mobilisations were held in practically every city and town. Participants in the protests held on the morning of the 12th—the largest since the years of the transition—exhibited the blue ribbon and whitewashed hands that had been used to symbolise opposition to ETA since the kidnapping of Julio Iglesias Zamora in 1993 and the murder of Federico Tomás y Valiente in 1996, respectively. Although the largest demonstrations were in Madrid and Barcelona, the most significant and emotional was the one held in Bilbao, the provincial capital of Biscay, which was headed by the prime minister, José María Aznar, and many of his ministers.

The ritual phase of this counter-performance stretched from the publication of ETA's communiqué on the afternoon of the 10th (ritual entry) until Blanco's funeral on Monday the 14th (ritual exit). Structurally, it centred around two liminal moments, the kidnapping and the murder, against which the collective mobilised, attaining a state of collective effervescence or communitas. The result of this communitas was that, in that moment, communities that had been divided over conflicting definitions of the national sacred fused, extolling the sacredness of the individual over any such divisions. This mutation of the hierarchical order of the sacred, in which priority is given to the sacredness of human life, restores the sacrificial nature of the victim, whose sacralisation forms the threshold for initiating a process of civil reparation of the cultural trauma experienced by the collective.

[13] Some time later, the press reported that, before ETA issued its communiqué, the Minister of the Interior's secretary received a threatening call, saying: "You bastards, you're going to pay for what happened with Ortega Lara. Long live the Free Basque Country!".

Of course, this communitas fades in time and the pre-existing order to which the divided communities pay allegiance is re-established. Nonetheless, as Turner has noted, once the ritual performance is over, reality has been transformed. Ultimately, this marked the beginning of the end for ETA.[14]

2.2. The 11-M Jihadist Attacks in Madrid

Islamic terrorism, says Alexander, is a post-political phenomenon representing the end of political possibilities, including both the most immediate possibilities, such as creating successful Arab states in the Middle East, and the more utopian possibilities, such as turning the *ummah* into a single great Moslem state. To this end, it seeks to create political and social instability by assassinating key leaders, sowing fear and obliging the authorities to adopt repressive measures that delegitimise its institutional network (Alexander 2017, pp. 200–1). Although Spain has suffered its consequences since the mid-1980s, until the beginning of the twenty-first century, the intelligence services did not believe it had the means to mount a terrorist attack on Spanish soil. Spain had become a jihadist target because of its participation in the Iraq War (Comisión de Investigación Sobre el 11 de Marzo de 2004 2005).

On 11 March 2004, Al-Qaeda carried out a series of bombings in which 192 people lost their lives and over 1800 were injured (Comisión de Investigación Sobre el 11 de Marzo de 2004 2005).[15] Ten bombs exploded during the morning rush hour on four trains heading towards Atocha station in Madrid, creating a scene of carnage, with hundreds of dismembered bodies strewn amongst the debris. Certain parallels can be traced between the collective reactions to the 9/11 attacks in the US and the 11-M bombings in Madrid. As in the case of 9/11 (Alexander 2017), the terrorist performance was initially received in triumph by the audience at which it was aimed. A report in The New Yorker related the satisfaction with which Al-Qaeda and its followers received news of the attacks:

> "An Al Qaeda statement posted on the Internet after the March 11th bombings declared, 'Being targeted by an enemy is what will wake us from our slumber'". (Wright 2004)
>
> A post on Facebook went further:
>
> On March 12th, the day after the train bombings, a message titled "The Goals of Al Qaeda in Attacking Madrid" had been posted by a writer calling himself Gallant Warrior. Echoing a theme that is frequently repeated on these sites, the writer noted that by carrying out its threat to Spain, Al Qaeda proved that its words were matched by actions: "Al Qaeda has sent a message to the crusading people: do not think that death and fear are only for the weak Muslims. ... Aznar, the American tail, has lost. And great fear has spread among the people of the countries in alliance with America. They will all be vanquished. Thank God for letting us live this long to see the jihad battalions in Europe. If anyone had predicted this three years ago, one would have said he was dreaming". (Wright 2004)
>
> And on ikhwan.net:
>
> A writer named Murad chastised those who condemned the Madrid bombings. "You pity the deaths of those non-Muslims so quickly! If Muslims had died in their lands in the manner the writer discusses, would he have cried for them?" A woman named Bint al-Dawa responded, "Brother Murad, Islam does not allow the killing of innocent people." A man who called himself "Salahuddeen2" entered the discussion: "We have said that we are against the killing of civilians anywhere, but the enemies of God kill Muslim civilians every day and do not feel shame. They should drink from the same bitter cup". (Wright 2004)

[14] Even the media changed the way in which they reported their attacks (Caminos et al. 2013).

[15] It is the second bloodiest attack in Europe after Pan Am flight 103, brought down over Lockerbie in 1988 (Pérez-Ventura 2014).

However, as in New York, the general public quickly mobilised. Thousands of Madrileños turned out to offer assistance to the victims at the scene of the attack, donating blood and giving whatever support they could. Anonymous heroes appeared, alongside the police, firefighters, health workers and psychologists, giving help to the victims. From that same day, ordinary citizens began a sort of pilgrimage to the site of the bombings, leaving all kinds of offerings to commemorate the victims. Altars were set up throughout the city, in the victims' workplaces, at iconic monuments and in businesses and shops. The citizenry took over the public space to sacralise it and commemorate the dead, but also to demand information and accountability (Ortiz García 2008).

A performance was being staged that rejected fear and which, in those initial moments, managed to fuse its audience, the entire country, in a community that unwaveringly espoused the sacred value of human life, intuiting a script that centred on a post-sacrificial narrative. However, unlike the aftermath of the 9/11 attacks, politicians and media commentators failed to provide a single interpretation of the facts that might create the script extolling the collective self-resisting adversity in the name of sacrosanct human life. This would have been an essential element for turning the performance that was already underway into a successful counter-performance in opposition to the performance of terror.

With general elections due to be held just three days after the attacks, the government, motivated by partisan interests, launched a campaign of disinformation, pointing the finger at ETA as the perpetrators of the crime, despite prima facie evidence to suggest that it had been committed by Islamic terrorists (Comisión de Investigación Sobre el 11 de Marzo de 2004 2005).[16] ETA's denials of involvement were scorned, as was a video issued on behalf of Al-Qaeda, claiming responsibility (Catalán 2005). Many Spanish media fell in line with the government position (Sampedro Blanco 2009). In the face of growing doubts among the public, the government resorted to tactics of intimidation, denouncing those who questioned their version as "contemptible". On 12 March, the government unilaterally called a demonstration against terrorism. The march in Madrid was fronted by a banner replicating one of the slogans used in the pre-electoral campaign, "Con las víctimas, con la Constitución y por la derrota del terrorismo" ("With the victims, with the constitution and for the defeat of terrorism"). The demonstrators, however, chanted back, "¿Quién ha sido?" ("Who was it?").

In short, the public were asked to choose between believing the government or believing ETA. Despite media efforts, the pre-existing climate of opinion did not favour the government's campaign.[17] As a result, the audience which in the initial performance had fused around the sacralisation of human life, split into two groups, those who accepted the sacrificial narrative of the government and those who rejected it, maintaining the post-sacrificial narrative. The media and politicians appealed to them, deploying the collective representations of the "Us" that divide Spanish society around its original trauma[18], unleashing a political struggle in which each side tried to present itself as a pure and sacred "Us" in contrast to the "Other", associated with the impure and contaminated.[19] Unlike the cases of the 9/11 attacks and that of Miguel Ángel Blanco, it did not manage to manufacture a successful counter-performance that could clearly prevail over the terrorist performance. And this partly achieved the goal of destabilising Spanish society, socially,

---

16 After the Madrid attacks, one of Prime Minister Aznar's advisers apparently said to him: "If it was ETA, we'll win by a landslide, but if it was the jihadists, the PSOE will win" (Ortega Dolz 2019).

17 The mass protests held the previous year against Spain's participation in the Iraq War, a decision taken unilaterally by Prime Minister Aznar, were still very present. Ninety-one percent of Spaniards were opposed to the war and 80% considered that it harmed the country (Centro de Investigaciones Sociológicas CIS 2003). In February 2003, this rejection had led over three million people to turn out in Madrid and Barcelona under the slogan "¡*No a la guerra*!" ("No War!"), the largest protests since the murder of Miguel Ángel Blanco (Martínez-Fornes 2003).

18 Divins (2016) also sees a relationship between the political conflict aroused by the interpretation of the attack and the Civil War, over whose memory a pact of silence had been imposed by the Transition.

19 One example was the accusations levelled against the Socialist Party by the People's Party for organising 20,000 people to protest in front of the PP's offices on the 13th, despite the fact that the event was a "flash" protest, spread by text message (Sampedro Blanco 2009).

politically and morally, intensifying the conflict on the definition of the sacred "Us" and the role played in it by the sacrificial sacralisation of the victims of terrorism.

3. Institutionalisation of the Sacrificial Sacralisation of the Victims

3.1. From Invisibility to Protagonism of the Sacrificed Victim

The institutionalisation of the sacrificial sacralisation of the victims that came with the 1999 Solidarity with the Victims of Terrorism Act (SVTA)[20] constitutes part of a process of civil reparation of the cultural trauma that had begun with the communitas that emerged following the kidnap and murder of Miguel Ángel Blanco. Before that date, there had been performances of solidarity with the victims, such as the mass turn-out for the funeral of the President of the Provincial Government of Gipuzkoa, Juan María Araluce, in 1976 and protests at the murder of the engineer José María Ryan in 1981. However, under the Franco dictatorship, anti-terrorist legislation included no provision favouring the victims (Serranò 2018), and during the transition, they received only very limited attention from the authorities.[21] The 9/11 attacks in the US and the 11-M attacks in Madrid drove changes to the legislation on victims, particularly the 2011 Recognition and Comprehensive Protection of Victims of Terrorism Act (RCPVTA).[22]

There was a shift from a period of invisibility to one in which the victims became the protagonists, whose suffering came to be considered as being a priority and a unifying force (Boutellier 2000). This epochal phenomenon also occurred in Spain (Mateo Santamaría 2018). The change was to affect social representations centring on the victims and the role played by their associations. In Spain, although the initial drive for recognition of the victims came from these organisations, growing support from the institutions meant that between 1988 and 2018, their numbers swelled from eight to thirty-eight, of which three were devoted to the victims of the 11-M attacks (Mateo Santamaría 2018). This reflects the diversity of outlooks with regard to the role of the victims and views on anti-terrorist policy.

The SVTA was presented as a way of honouring and recognising the sacrifice of the victims of terrorism. The law defined the victims as those people who, in sacrificing their life or their freedom, exhibited and defended the values of society. Given that, at the time it was passed into law, the category of "victim" was colonised by the victims of ETA, they were positioned in the sacred ambit by the representations which the state began to institutionalise through laws and performances. At the same time, in the debate on the RCPVTA, some parties on the left suggested that the victims' sacrifice for the benefit of society should be considered as involuntary (Amendment # 52). However, this amendment was not passed.

According to the RCPVTA, the victim is a symbol; it states that by attacking the victim, the aim is to attack the values of society and the rule of law, and thus confers a political significance on the victim. The victim emerges as a sacrificial offering by means of which it is hoped to achieve the emotional adhesion of the community to the institutions, reinforcing their link with the nation-state. In the absence of heroes or martyrs who had self-immolated in the name of the nation, their place is taken by the victims of terrorism.[23] Thus, the institutional narrative makes an effort to present them as sacrificially sacralised victims.

This representation, elaborated a posteriori, should be successful, if we bear in mind that in secularised societies with a fragmented morality, the victim becomes a moral refer-

20 *Ley 32/1999 de Solidaridad con las Víctimas del Terrorismo.*

21 Royal Decree Law 3/1979 on protection of the public security in which the state assumed responsibility for issuing compensation to the victims.

22 *Ley 29/2011 de Reconocimiento y Protección Integral a las Víctimas del Terrorismo.*

23 As stated above, according to Hénaff, victims of terrorist attacks are not always sacrificial. Furthermore, strictly speaking, a distinction must be made between martyr, national hero and suicide bomber, as they represent different historical manifestations of the post-sacrificial narrative. As Beriáin (2007) indicates, the martyrs embody the post-sacrifice narrative introduced by Christianity: they are victims who know that they are going to die and who choose to do so by endowing their death with a solidarity meaning with a discriminated group. In the process of modern secularization, the national hero takes the martyr's place, so the duty of dying for the country provides meaning to his potential death. The suicide bomber rises in late or global modernity as a reinvention of the historical archetype of the martyr, signifying both the duty to die for the *ummah* and the rejection of Western modernity.

ence such that "Durkheim's organic solidarity has dissolved into a victimalised solidarity" (Boutellier 2000). When there is a consensus as to what the victim represents, the social legitimacy that the victim enjoys extends to the institutions that defend him or her, and each mutually feed of one another. However, in Spain, this aspect proved problematic, because the institutional narrative on the victims of terrorism clashed with the sacrificial narratives that established the hierarchical orders of the national sacred in conflict.

3.2. *The Delimitation of the Pure and the Impure*

The new context of international terrorism that emerged following the 9/11 attacks in the US facilitated the adoption of certain political measures in Spain, which could be defended as being part of the cause of the victims of terrorism. Prime Minister Aznar seized the opportunity to declare that "all terrorisms" were alike, equating ETA with Bin Laden. He tried to force through the illegalisation of Batasuna, the radical left-wing Basque nationalist party, which the government considered to act as a support for ETA (Anasagasti 2007). He managed to establish new legal and symbolic limits with regard to the pure and the impure that took concrete form in the 2002 Political Parties Act[24], which drew a distinction between those organisations that acted with respect for democracy and those that based their political action on collusion with terror, violence and the violation of human rights. The enactment of the law led to the banning of Batasuna, judicial persecution of its milieu and the stigmatisation of anyone who opposed the concept of democracy extolled in the Act. From that moment on, there was an intensification in the tabooisation of the terrorist, which was considered to contaminate anyone coming into contact with it (Zulaika and Douglass 1996).[25]

An order was imposed, along the lines established by Douglas (1991), which redefined "the Us" by tracing the outlines of all that was clearly out of place and threatened that just order, representing the danger, the dirty and the contaminating. An attempt was made to unite the community in protecting itself from a common threat, under pain of contamination and moral opprobrium. "The Us" was thus redefined with the victims of terrorism of ETA at its core. Violating the taboo not only meant being excluded but also calling into question the sacrifice of the victims, something which could not be tolerated in a sacrificial narrative built on the logic of confrontation.

The 11-M attacks extended the map of terrorist victims and perpetrators. However, the government tried to impose the representation of the pure and the impure that it had constructed in reference to ETA terrorism. Identifying ETA as the perpetrator of the 11-M attacks was presented as a moral conviction. Aznar could have drawn a likeness between the two terrorisms, as he had before, but the events demonstrated a difference in nuances between "old" and "new" terrorism. These nuances introduced ambivalence, ambiguity and indecisiveness into a representation based until then on a clear dichotomy. The government even went so far as to suggest that the attacks had been organised jointly by ETA and Al-Qaeda (Catalán 2005). Another sacred element had to be invoked to re-establish the binary logic. Despite the fact that national reconciliation is enshrined in the Spanish constitution, a distinction was drawn between "constitutionalists and non-constitutionalists", the latter being relegated to the realm of the impure.

The imposition of this rigid binary logic hindered any consensus as to those whom the RCPVTA should recognise as victims; victims of state abuse, victims of the Franco dictatorship and victims of other terrorist organisations with political objectives were all excluded, despite opposition from nationalist and left-wing parties. For this reason, no mention is made of any specific group of victims. Moreover, the law considered democratic participation to be incompatible with organisations that represent or justify terrorism. In short, any flexibilisation in the limits set out (such as the inclusion of other victims) was

---

24 *Ley Orgánica 6/2002 de Partidos Políticos.*

25 This taboo remains operative to this day. In 2018, for example, Alfonso Sánchez, President of the Association of Victims of Terrorism (AVT), was removed from his post for holding institutional encounters with leader of the Basque nationalist left (Ballesteros 2018).

interpreted as a betrayal of "the" victims of terrorism that would signify whitewashing the perpetrator or imposing a "falsified" memory.

3.3. Decorations as Functional Substitutes for Purification Rituals

Decorations are a response to victims' need for a civil recognition or reparation, associated with remembrance and commemoration. In their process of resilience, the victims must be able to see that it is they and not the terrorists who are at the heart of society's attention (Ivankovic et al. 2017). The SVTA created the Royal Order of Civil Acknowledgement of Victims of Terrorism[26] as a recognition of their sacrifice The order's Grand Cross (*Gran Cruz*) was awarded posthumously to those killed in terrorist attacks and a Commendation (*Encomienda*) to those who had been injured or kidnapped. With the amendment to this law in 2003, a further step was taken towards a legislative delimitation of the pure and the impure as applied in Spain. Victims were required to have an untainted personal background to be eligible for decoration. The values set out in the Constitution, human rights and laws on victims of terrorism were added to the realm of the pure. This criterion was maintained in the RCPVTA, which moreover added that the victims' social recognition derived from their political significance and it was held up as a tool for the ethical, social and political delegitimisation of terrorism.

The inclusion of new requirements for receiving decorations is probably related to an institutional concern that some victim of a terrorist act might not "really" represent the social values generically attributed to him/her. The public and individualised recognition of victims who do not represent those values would constitute a transgression of the established limits, hence the need to guarantee their purity. The decorations become functional substitutes for purification rituals, necessary in order to declare the victim's innocence, which is no longer presumed solely by virtue of the unjust violence suffered. In the ceremonies at which the decorations are awarded, there is a ritual remembering and the sacrificial sacralisation involves a process whereby the institutional structure guarantees that the victim effectively embodies values that it wishes to prevail and considers to be sacred.

Thus, the public conferrals of the medals become performances in which, at the same time as the victim is exalted, the values of the regime are reinforced. The occasion is commonly used to delegitimise—with the collusion of the self-styled constitutionalist parties—those who reject the victors-and-vanquished logic and call for recognition of all victims. In this context, the distinction between victims who deserve decoration and those who do not is enshrined in a dynamic of confrontation: only the victim who can function as a scapegoat, freeing the group from all blame, is worthy of civil recognition. Paradoxically, the distinction between victims prevents the entire community from coming together and fusing around the sacrificially sacralised victim; indeed, it actually foments division. Even victims of the 11-M attacks who have received institutional recognition feel excluded by this logic. One such individual, Antonio Miguel Utrera (Fundación Víctimas del Terrorismo 2020), called on the authorities in the following terms: "I ask you to accompany and defend us from those who brandish the flag of hatred and attack us. However, I ask you not to accompany or support us if our claims are marked by the hatred or sense of victimisation that leads to revenge".

3.4. Tributes to the Victims in Divided Communities

In Spain, tributes to victims of terrorism have not always fomented unity. Different performances of solidarity with victims have also served communities with different identitary references to reaffirm their concept of the "Us". One example of this dynamic can be seen in the acts scheduled in parallel in two Basque locations in June 1995. One section of Basque society went to the "concert for peace" to demand that ETA release the engineer José María Aldaya, kidnapped some months after the killing of councillor

[26] *Real Orden de Reconocimiento Civil a las Víctimas del Terrorismo.*

Gregorio Ordoñez (Barbería 1995) Another sector, sympathetic to Batasuna, expressed its indignation at the way in which the bodies of ETA militants, Lasa and Zabala, murdered in 1983 by GAL, had been interred (Guenaga 1995). Some of the speakers at the concert called for an end to violence in the Basque Country and argued that certain judicial and police actions repel the general public. In the tribute to Lasa and Zabala, there were chants against those who wore blue ribbons and a banner with the face of the two young men, stating "you cannot kill the soul of a people", in an effort to identify their victims with those of all of Basque society.

This division began to be weakened following the communitas that arose out of the kidnapping and killing of Miguel Ángel Blanco. In order to maintain the spirit of communion, the Aznar government sought to institutionalise it through performances, such as the concert held in Madrid in September 1997 to pay homage to Blanco, which was broadcast on state television. Under the slogan "United for Peace", artists representing different identities and political persuasions took part in the act. One of the acts was Valencian singer-songwriter Raimon, who was introduced as a Catalan who deserved thanks for inspiring people to learn other languages and to know, respect and love a nation's signs of identity. Raimon presented his performance as a contribution to the "fight against death", choosing the song "País Vasc" (Basque Country), which he had written during the years of the dictatorship. The song, a tribute to the suffering of the Basque people, was sung in the Valencian language, and the singer-songwriter took the opportunity to remind his audience that the language had been banned several times during the Franco dictatorship.[27] This earned him booing from a section of the crowd, which attempted to disrupt the performance.[28] The incident appeared to show that the fusion experienced as a communitas had been short-lived and that it would not be simple to overcome the history of divisions in which a given interpretation of the "unity" of Spain has been sacralised above other pluralist forms of viewing coexistence.

It is clear to see the binary logic that has been extended to other social representations that end up relating the domains of the pure and the impure with other—also sacralised—notions. The different nationalisms that coexist within Spain (Spanish, Basque, Catalan and Galician) and the identitary issues linked to them are juxtaposed, when they are not combined, with aspects of a different nature, such as those related to solidarity, empathy and identification with the suffering of the victims of terrorism.

For their part, the institutions continued to step up performances of recognition of the victims in a continuous attempt to appeal to that unity surrounding them. The RCPVTA meant greater institutionalisation of those acts, providing support for victims to participate in all institutional acts that affected them. It also declared 27 June as a day of remembrance and homage to the victims and 11 March as the commemoration of the European Day of Victims of Terrorism. Other forms of recognition and remembrance of the victims included acts, symbols, monuments and similar elements.

These recognitions were linked to the effective reparation and the core ideas of the new legislation were defined as being "remembrance, truth, justice and dignity". However, these ideas were also framed in the discourse of victor/vanquished, in which full reparation to the victim involves a firm anti-terrorist policy that brooks no form of complicity. Despite the apparent clarity of this approach, it is nonetheless problematic when, for example, the different victims' associations (or the victims considered individually) struggle to discover "the truth" or interpret the overcoming of their trauma in different ways.

The separate tributes of different victims' associations and political parties fed the division, transmitting that split to society as a whole. This division increases the risk that the cause of the victims may be politicised by a political party or by associations of victims

27 The recognition of Spain's linguistic plurality is the subject of political dispute. A structured review of linguistic repression can be found in Torrealdai (1998).

28 El dia que Raimon va ser xiulat per cantar en valencià (https://www.youtube.com/watch?v=NBAF-dihiko&feature=youtu.be) (accessed on 20 November 2020).

that want to impose their own anti-terrorist policy. This weakens the cause of the victims, since doing justice through reparation for the personal and social damage caused requires both social recognition of the victims and a healing of the social fracture caused by the crime (Mate 2008).

The commemoration of the 11 March attacks is an example of the existing social and political division. Year after year, two of the leading victims' associations, Asociación 11M Víctimas and the AVT (Association of Victims of Terrorism), stage separate performances in Madrid, which are attended by different political representatives. Both organisations continue to demand that the full truth be made known, but in opposing senses. While the former recalls the untruths of Aznar, the second insinuates the existence of hidden police reports that would provide a different version of the events (Clemente 2017).

The political parties also foster the division. This can be seen in the Day of Remembrance in the Basque Country for recognition of the victims: "Since the day was first commemorated in 2010, it has been marked by divisions as to the significance of the date and the category of victims" (Lecumberri 2018).[29] So, for example, there have been years in which left-wing Basque nationalist parties did not attend because the tribute did not include the victims of all forms of violence. However, when the victims of politically motivated violence and police abuse were included, the People's Party did not attend.

The institutional recognition of the victims of terrorism contrasts with the public welcoming ceremonies for ETA members released from prison, appearing to play out a competition for the performative recognition of sacrificed heroes The AVT has denounced the fact that this type of "tribute" offends and humiliates them and have even taken the matter to the courts (Gorospe 2019). Arnaldo Otegi, leader of the Basque nationalist left, answers that "they do not do it humiliate anyone but to welcome a person who is returning to his town" (Izarra 2019) and that the problem lies in the interpretation given by others to such receptions. Insofar as the receptions for ETA members revive the trauma experienced by their victims, they reinforce the binaries that categorise ETA in the realm of the impure; it also reinforces the very survival of that binary logic that is opposed to the emergence of a more integrating collective memory.

In short, the sacrificial sacralisation of the victim also extends to the sacralisation of his or her memory, whose purity is jealously guarded so that the sacralised and heroic victim is not sullied. The victims were the object of victimisation, but all the institutional narrative a posteriori, accompanied by certain performative acts, portrays them as heroic victims. This sacrificial commemoration is reminiscent of a state in war which requires an identifiable enemy, but which today not only faces less distinct threats (such as jihadist terrorism) but increasingly sees itself impelled to incorporate a post-sacrificial logic in which the sacrality of the person is imposed over any other hierarchical order of the sacred.

4. Conclusions

The killing of Miguel Ángel Blanco and the 11-M attacks questioned society by transforming the way in which it addressed terrorism. Insofar as the suffering of the victims became a collective suffering, a society that was divided over the memory of the original trauma and the trauma of terrorism achieved moments of fusion. Here, one can see the role of the performance in the process of the victims' sacralisation and also the prevalence of a post-sacrificial narrative in the communitas that had arisen in the moment of social effervescence. However, neither the crime against Blanco, nor 11-M, nor even the dissolution of ETA have, for the time being, managed to make that fusion permanent. Despite the transformation that has occurred, when the communitas is dissolved and there is a return to the institutionalised structure, the sacrificial logic re-emerges. The sacrificial and post-sacrificial narratives coexist, conditioned by the fact that Spain is a divided society, sustained on binary and exclusive representations. This is projected in the social represen-

29 The date of the celebration has great significance: "The 10 November, the only day in the calendar on which there have been no victims of terrorist acts" (Lecumberri 2018).

tations of an ideal order in which the victim of terrorism, subjected to functional substitutes of purification established by the institutions, becomes the focal point in the delimitation of the pure and the impure. These narratives become caught up in the dispute over the definition of the "Us". And as long as this does not find a channel through which it can be resolved, it will indeed be difficult to reach a consensus that can overcome the ambivalence with regard to the two narratives, in such a way that the anti-sacrificial logic prevails.

Finally, this type of analysis—the narratives on the victims—could also be applied to other contexts. For instance, exploring and establishing connections with the process of victims' identity recovery and construction in dictatorships and violent conflicts like the ones in Latin America or in post-Fascist Italy.

Author Contributions: Conceptualisation, E.A. and J.M.P.-A.; methodology, E.A. and J.M.P.-A.; formal analysis, E.A. and J.M.P.-A.; investigation, E.A. and J.M.P.-A.; resources, E.A. and J.M.P.-A.; writing—original draft preparation, E.A. and J.M.P.-A.; writing—review and editing, E.A. and J.M.P.-A. All authors have read and agreed to the published version of the manuscript.

Funding: This research received no external funding.

Acknowledgments: To the I-Communitas Institute of the Public University of Navarra for funding of translation.

Conflicts of Interest: The authors declare no conflict of interest.

References

Alexander, Jeffrey C. 2004. Toward a Theory of Cultural Trauma. In *Cultural Trauma and Collective Identity*. Edited by Jeffrey C. Alexander, Ron Eyerman, Bernard Giesen, Neil J. Smelser and Piotr Sztompka. Berkeley: University of California Press, pp. 1–30.

Alexander, Jeffrey. 2006. Social performance between ritual and strategy. In *Social Performance.Symbolic Action and Cultural Pragmatics*. Edited by Jeffrey C. Alexander, Bernhard Giesen and Jason L. Mast. Cambridge: Cambridge University Press, pp. 29–90.

Alexander, Jeffrey. 2017. *Poder y Performance*. Madrid: CIS.

Anasagasti, Iñaki. 2007. De cómo nos Opusimos a la Ley de Partidos y Cómo nos Quedamos Más Solos Que la Una. *EAJ-PNV*. Available online: https://www.eaj-pnv.eus/es/documentos/6336/de-como-nos-opusimos-a-la-ley-de-partidos-y-como-n (accessed on 9 October 2020).

Ballesteros, Roberto. 2018. Terremoto en la AVT: Las Víctimas 'Echan' al Presidente por Reunirse con 'Abertzales'. *El Confidencial*. Available online: https://www.elconfidencial.com/espana/2018-04-05/avt-victimas-terrorismo-alfonso-sanchez_1544670/ (accessed on 9 October 2020).

Barbería, José L. 1995. 18.000 Personas Reclaman en el Concierto por la Paz el fin de la Violencia en Euskadi. *El País*. Available online: https://elpais.com/diario/1995/06/25/espana/804031210_850215.html (accessed on 9 October 2020).

Beriáin, Josetxo. 2007. Chivo expiatorio-mártir, héroe nacional y suicida-bomba: Las metamorfosis sin fin de la violencia colectiva. *Papers* 84: 99–128. [CrossRef]

Beriáin, Josetxo. 2016. Formas modernas de resacralización en disputa. *La nación y la persona. Revista Internacional de Sociología* 74: 1–13.

Beriáin, Josetxo. 2017. Las metamorfosis del don: Ofrenda, sacrificio, gracia, substituto técnico de Dios y vida regalada. *Política y Sociedad* 54: 641–63.

Beriáin, Josetxo. 2021. The Endless Metamorphoses of Sacrifice and its clashing narratives. *Religions* 11: 684. [CrossRef]

Blanco, José M., and Jéssica Cohen. 2016. Viejo y Nuevo Terrorismo. *Enfoque* 2. Centro de Análisis y Prospectiva. Gabinete Técnico de la Guardia Civil. Available online: https://intranet.bibliotecasgc.bage.es/intranettmpl/prog/local_repository/documents/17873.pdf (accessed on 15 April 2020).

Boutellier, Hans. 2000. *Crime and Morality: The Significance of Criminal Justice in Post-Modern*. Dordrecht: Kluwer Academic Publisher.

Caminos, José M., José Ignacio Armentia Vizuete, and Flora Marín Murillo. 2013. El asesinato de Miguel Ángel Blanco como ejemplo de *key event* en el tratamiento mediático de los atentados mortales de ETA. *adComunica. Revista de Estrategias, Tendencias e Innovación en Comunicación* 6: 139–60. Available online: http://repositori.uji.es/xmlui/bitstream/handle/10234/78366/130-332-1-PB.pdf?sequence=1&isAllowed=y (accessed on 15 April 2020).

Catalán, Miguel. 2005. Prensa, Verdad y Terrorismo: La Lección Política del 14-M». *El Argonauta Español* 2. Available online: http://journals.openedition.org/argonauta/1191 (accessed on 4 November 2020). [CrossRef]

Centro de Investigaciones Sociológicas CIS. 2003. Barómetro de Febrero 2003. Estudio nº 2.481. Available online: http://www.cis.es/cis/export/sites/default/-Archivos/Marginales/2480_2499/2481/Es2481.pdf (accessed on 20 October 2020).

Clemente, Enrique. 2017. Las Víctimas del 11M Exigen Trece Años Después Que se Sepa Toda la Verdad. *La Voz de Galicia*. Available online: https://www.lavozdegalicia.es/noticia/espana/2017/03/12/victimas-11m-exigen-trece-anos-despues-sepa-verdad/0003_201703G12P18991.htm (accessed on 20 October 2020).

Comisión de Investigación Sobre el 11 de Marzo de 2004. 2005. Texto Aprobado por el Pleno del Congreso de los Diputados, en su Sesión del Día 30 de Junio de 2005, Resultante del Dictamen de la Comisión de Investigación Sobre el 11 de Marzo de 2004 y de los Votos Particulares Incorporados al Mismo. Boletín Oficial de las Cortes Generales, VIII Legislatura, nº 424. Available online: http://www.congreso.es/public_oficiales/L8/CONG/BOCG/D/D_242.PDF (accessed on 6 September 2020).

De la Calle, Luis, and Sánchez-Cuenca Ignacio. 2004. La selección de víctimas en ETA. *Revista Española de Ciencia Política* 10: 53–79.

De Pablo, Santiago. 2018. Julio de 1959: El nacimiento de ETA. *Historia Actual Online* 48: 45–59. Available online: https://historia-actual.org/Publicaciones/index.php/hao/article/view/1689 (accessed on 15 November 2020).

Dingley, James, and Michael Kirk-Smith. 2002. Symbolism and Sacrifice in Terrorism. *Small Wars & Insurgencies* 13: 102–28.

Divins, Thomas J. 2016. Four Days that Transformed Spain 11-M Impact on Memory Recovery Examined through the Lens of Duality. Master's thesis, West Virginia University, Morgantown, VA, USA. Available online: https://researchrepository.wvu.edu/etd/5501 (accessed on 30 November 2020). [CrossRef]

Douglas, Mary. 1991. *Pureza y Peligro: Un Análisis de los Conceptos de Contaminación y Tabú*. Madrid: Siglo XXI.

Durkheim, Émile. 1973. Individualism and the Intellectuals. In *Emile Durkheim on Morality and Society*. Edited by Robert Bellah. Chicago: University of Chicago Press, pp. 43–57.

Eliade, Mircea. 1985. *Lo sagrado y lo Profano*. Barcelona: Editorial Labor.

Fundación Víctimas del Terrorismo. 2020. Homenaje del Ministerio del Interior a las Víctimas del Terrorismo. *Revista de la Fundación Víctimas del Terrorismo* 70: 23. Available online: http://fundacionvt.org/wp-content/uploads/2020/05/fvt70.pdf (accessed on 9 October 2020).

Funes, María J. 1998. *La Salida del Silencio. Movilizaciones por la paz en Euskadi 1986–1998*. Madrid: Akal.

Giesen, Bernhard. 2006. Performing the sacred: A Durkheimian perspective on the performative turn in the social sciences. In *Social Performance, Symbolic Action and Cultural Pragmatics*. Edited by Jeffrey C. Alexander, Bernhard Giesen and Jason L. Mast. Cambridge: Cambridge University Press, pp. 325–67.

Girard, René. 2005. *La Violencia y lo Sagrado*. Barcelona: Anagrama.

González, Juan Manuel González. 2012. La violencia política de la extrema derecha durante la Transición española (1975–1982). Coetánea. In *Actas del III Congreso Internacional de Historia de Nuestro Tiempo*. Edited by Carlos Navajas Zubeldia and Diego Iturriaga Barco. Logroño: Universidad de La Rioja, pp. 365–76.

Gorospe, Pedro. 2018. Teoría y Práctica de la Socialización del Sufrimiento. *El País*. Available online: https://elpais.com/politica/2018/04/28/actualidad/1524913595_360804.html (accessed on 20 October 2020).

Gorospe, Pedro. 2019. El Parlamento Vasco Rechaza los Recibimientos a Presos de ETA que Apoya la Izquierda 'Abertzale'. *El País*. Available online: https://elpais.com/politica/2019/10/03/actualidad/1570092352_863926.html (accessed on 20 October 2020).

Guenaga, Aitor. 1995. Homenaje a Lasa y Zabala en Tolosa Tras una Jornada de Lucha con 10 Detenidos. *El País*. Available online: https://elpais.com/diario/1995/06/25/espana/804031211_850215.html (accessed on 20 October 2020).

Gurruchaga, Ander. 1988. *El código Nacionalista vasco Durante el Franquismo*. Barcelona: Anthropos.

Hénaff, Marcel. 2002. *The Price of Truth. Gift, Money and Philosophy*. Stanford: Stanford University Press.

Hubert, Henri, and Marcel Mauss. 1899. Essai sur la Nature et la Fonction du Sacrifice. *Année sociologique* tome II: 29–138. Available online: http://classiques.uqac.ca/classiques/mauss_marcel/melanges_hist_religions/t2_sacrifice/Melanges_2_sacrifice.pdf (accessed on 15 April 2020). [CrossRef]

Ivankovic, Aleksandra, Levent Altan, and An Verelst. 2017. *How Can the EU and the Member States Better Help Victim of Terrorism?* Brussels: Policy Department for Citizens' Rights and Constitutional Affairs.

Izarra, Josean. 2019. Arnaldo Otegi Anuncia Que "Habrá 250 Recibimientos Más a Presos". *El Mundo*. Available online: https://www.elmundo.es/pais-vasco/2019/08/02/5d4406e0fc6c83ec168b464f.html (accessed on 20 October 2020).

Izquierdo, Jesús. 2017. Memoria normalizada: 1936 en la España de la impunidad. *Rey Desnudo Año VI* 11: 119–34.

Joas, Hans. 2019. *La sacralidad de la Persona. Una Nueva Genealogía de los Derechos Humanos*. Buenos Aires: UNSAM EDITA.

Lecumberri, Jokin. 2018. La Unanimidad Imposible del Día de la Memoria en Euskadi. *La Vanguardia*. Available online: https://www.lavanguardia.com/local/paisvasco/20181110/452801301467/dia-de-la-memoria-pais-vasco-unanimidad-terrorismo-eta-violencia-victimas.html (accessed on 20 October 2020).

Martínez-Fornes, Almudena. 2003. El «No a la Guerra» Reúne la Mayor Protesta Desde el Asesinato de Miguel Ángel Blanco. *ABC*. Available online: https://www.abc.es/internacional/abci-no-guerra-reuneynbsp-mayor-protesta-desde-asesinato-miguel-angel-blanco-200302160300-162383_noticia.html (accessed on 20 October 2020).

Mate, Reyes. 2008. *Justicia de las Víctimas. Terrorismo, Memoria y Reconciliación*. España: Anthropos.

Mateo Santamaría, Eduardo. 2018. La contribución del movimiento asociativo y fundacional a la visibilidad de las víctimas del terrorismo en España. *Revista de Victimología* 7: 9–46. [CrossRef]

Ortega Dolz, Patricia. 2019. El Gobierno de Aznar me Pidió que Asumiera su Mentira Sobre el 11-M. *El País*. Available online: https://elpais.com/politica/2019/03/10/actualidad/1552221291_945279.html (accessed on 20 October 2020).

Ortiz García, Carmen. 2008. Memoriales del atentado del 11 de marzo en Madrid. *Cadernos de Estudos Africanos* 15: 47–61. [CrossRef]

Pérez-Agote, Alfonso. 1984. *La Reproducción del Nacionalismo: El Caso Vasco*. Madrid: Siglo XXI.

Pérez-Ventura, Óscar. 2014. La amenaza de Al-Qaeda en España diez años después del 11-m. *Revista Aequitas* 4: 389–409.

Preston, Paul. 2012. *The Spanish Holocaust: Inquisition and Extermination in Twentieth-Century Spain*. London: HarperPress.

Sabucedo, José Manuel, Mauro Rodríguez, and W. López López. 2000. Movilización social contra la violencia política: Sus determinantes. *Revista Latinoamericana de Psicología* 32: 345–59.

Sampedro Blanco, Víctor. 2009. Conspiración y pseudocracia. O la esfera pública a cinco años del colapso del 11-M. *Viento Sur* 103: 60–68. Available online: https://cdn.vientosur.info/VScompletos/Sampedro.pdf (accessed on 20 October 2020).

Schmid, Alex P. 2011. *The Routledge Handbook of Terrorism Research*. New York: Routledge Handbooks, Taylor and Francis.

Serranò, Agata. 2018. *Las víctimas del Terrorismo: De la Invisibilidad a los Derechos*. Madrid: Thomson Reuters Aranzadi.

Tejerina, Benjamín. 2015. Nacionalismo, violencia y movilización social en el País Vasco. Factores y mecanismos del auge y declive de ETA. *Papeles del CEIC* 3: 136. [CrossRef]

Tognato, Carlo. 2013. Extending Trauma Across Cultural Divide: On Kidnapping and Solidarity in Colombia. In *Narrating Trauma: On the Impact of Colective Suffering*. Edited by Ron Eyerman, Jeffrey C. Alexander and Elizabeth Butler Breese. Boulder and London: Paradigm Publishers.

Torrealdai, Joan Mari. 1998. *El Libro negro del euskera*. Donostia: Ttarttalo.

Turner, Victor M. 1969. *The Ritual Process: Structure and Anti-Structure*. Chicago: Aldine Publishing Company.

Woodworth, Paddy. 2001. *Dirty War, Clean Hands: ETA, the GAL and Spanish Democracy*. Cork: Cork University Press.

Wright, Lawrence. 2004. The Terror Web. Were the Madrid bombings part of a new, far-reaching jihad being plotted on the Internet? *The New Yorker*. April 2. Available online: https://www.newyorker.com/magazine/2004/08/02/the-terror-web (accessed on 15 November 2020).

Zulaika, Joseba. 1991. Reyes, políticos, terroristas: La función ritual de ETA en relación al nacionalismo vasco. *Revista de Antropología Social* 0: 217–30.

Zulaika, Joseba, and William A. Douglass. 1990. On the Interpretation of Terrorist Violence: ETA and the Basque Political Process. *Comparative Studies in Society and History* 32: 238–57.

Zulaika, Joseba, and William A. Douglass. 1996. *Terror and Taboo. The Folies, Fables and Faces of Terrorism*. New York and London: Routledge.

Zulaika, Joseba, and William A. Douglass. 2008. The terrorist subject: Terrorism studies and the absent subjectivity. *Critical Studies on Terrorism* 1: 27–36. [CrossRef]

MDPI
St. Alban-Anlage 66
4052 Basel
Switzerland
Tel. +41 61 683 77 34
Fax +41 61 302 89 18
www.mdpi.com

Religions Editorial Office
E-mail: religions@mdpi.com
www.mdpi.com/journal/religions